D0208449

The Space Environment

The Space Environment
Implications for Spacecraft Design

Alan C. Tribble

Princeton University Press ■ Princeton New Jersey

Published by Princeton University Press, 41 William Street,
Princeton, New Jersey 08540
In the United Kingdom: Princeton University Press,
Chichester, West Sussex

Library of Congress Cataloging-in-Publication Data

Tribble, Alan C., 1961–
The space environment : implications for spacecraft design / Alan C. Tribble.
p. cm.
Includes index.
ISBN 0-691-03454-0
1. Space environment. 2. Space vehicles—Design and construction. I. Title
TL1489.T75 1995
629.47'1—dc20 94-46481

This book has been composed in Times Roman

The publisher would like to acknowledge the author of this volume for providing the camera-ready copy from which this book was printed

Text design provided by Carmina Alvarez

Princeton University Press books are printed on acid-free paper and meet the guidelines for permanence and durability of the Committee on Production Guidelines for Book Longevity of the Council on Library Resources

Printed in the United States of America

10 9 8 7 6 5 4 3 2 1

To Matthew,

who observes the Moon with a sense of innocence and awe
that reminds me of the exhilarating feeling I experienced late
one special evening in July 1969 which started me on this
memorable journey

■ Contents

Preface and Acknowledgments

This book introduces the basic physics used to describe the space environment and illustrates how the environment interacts with spacecraft engineering subsystems and payloads. In this sense, it provides a bridge between the fields of space physics and astronautical engineering. The book is intended to serve as a basic reference for spacecraft system and payload engineers and is also suitable for use as a text in a one-semester introductory course on space environment effects. The level of detail provided is sufficient to illustrate the significance of the more important concepts, while voluminous detail on the nature of specific piece-part interactions or available computer modeling techniques has been intentionally avoided.

Numerous individuals have assisted in the preparation of this manuscript and deserve special mention. Ron Lukins and Eric Watts of Rockwell International's Space Environment Effects Laboratory are especially commended for their assistance in providing data, reviewing the contents, and offering encouragement when the task seemed hopeless. Jim Haffner of Rockwell International, William Metzger of Martin Marietta, and Rob Suggs of NASA Johnson Space Center served as technical reviewers and helped ensure a quality final product. The AE585 students at the University of Southern California, who field tested a draft copy of the manuscript in the spring of 1994, are applauded for their patience and thoughtfulness in helping to smooth the rough edges.

Last, but certainly not least, I cannot underestimate the immeasurable contributions made by my family. My wife and best friend Beth provided an unending supply of love, understanding, and moral support throughout the months required by this undertaking. My son Matthew provided enjoyable diversions during long nights at the keyboard, and our cats offered assurance that I could perform a useful function simply by providing a lap for them to sleep in. They are all to be commended for their help.

I will spare the reader my horror stories of the magnitude of the work that was involved to develop the material to this point lest I suppress the "publish or perish" gene in prospective authors and deprive the world of the fruits of their labors. Suffice it to say that it is now behind me, I have survived, and I can say, "This, with the help of God, I have done."

Long Beach, California
March 1995

The Space Environment

1 Introduction

In the beginning God created the heaven and the Earth.
—Genesis 1.1

1.1 Overview

If there is one common thread connecting modern man to his primitive ancestors it must surely be the fact that every individual on this planet has, at some time or another, lain awake at night staring at the heavens and pondering the nature of the universe. Since the beginning of time humans have looked at the stars, the planets, and the Moon and dreamed of traveling among them. Isaac Newton had a complete understanding of what was required to place an object in orbit around the Earth over 250 years ago. However, it was 1957 before mankind developed the technological ability to leave the Earth's surface and take the first tentative steps in the exploration of our solar system. Space travel is hampered not only by the difficulty of getting a spacecraft into orbit, but also by the fact that spacecraft must be designed to operate in environments that are quite different from those found on the Earth's surface. Although it is often considered a near perfect vacuum, the type of "space" that is encountered by an orbiting spacecraft may contain significant amounts of neutral molecules, charged particles, micron-sized particulates, and electromagnetic radiation. Each of these environments has the potential to cause severe interactions with spacecraft surfaces or subsystems and may, if not anticipated, severely impact mission effectiveness. The National Geophysical Data Center in Boulder, Colorado, has compiled a database of 2779 spacecraft anomalies from 1971 to 1989 that were related to interactions with the space environment. Studies of past NASA and Air Force spacecraft indicate that approximately 20–25% of all spacecraft failures are related to interactions with the space environment. Some key examples of spacecraft anomalies are listed in table 1.1.

Table 1.1

A Partial List of Spacecraft Anomalies

Spacecraft	Anomaly
Anik E-1 and E-2	• Failure of momentum wheel control systems during spacecraft charging event
Ariel 1	• Failed following detonation of high-altitude atomic tests
Geostationary Operational Environment Satellite (GOES)	• Numerous phantom command anomalies related to arc discharges from surface charging
Global Positioning System (GPS)	• Evidence of photochemically deposited contamination on solar arrays decreasing power output • Degradation of thermal control surfaces
Intelsat K	• Command anomalies related to arc discharges from surface charging
Long Duration Exposure Facility (LDEF)	• One month from atmospheric reentry when retrieved • Numerous MMOD impacts • Extensive contamination and AO degradation • Induced radiation
Pioneer Venus	• Several command memory anomalies related to high-energy cosmic rays
Skylab	• Reentered atmosphere as the result of increased atmospheric drag
Space Shuttle	• Numerous micrometeoroid/debris impacts • Shuttle glow • Collision avoidance maneuvers
Ulysses	• Failed during peak of Perseid meteoroid shower

Because it is the study of how the space environment effects a spacecraft, the field of study that specializes in the investigation of these interactions is called *space environment effects*. In 1993 the National Aeronautics and Space Administration (NASA) recognized the importance of the field of space environment effects by forming a national program to coordinate efforts in this area. In the same year, the International Standards Organization (ISO), under charter from the United Nations, formed a space systems technical committee which will work, among other things, to develop internationally recognized space environment standards. In 1994 NASA issued the first space environment effects research announcement and published orbital environment guidelines for spacecraft development.[1]

This text seeks to bridge the gap between space physics and astronautical engineering by presenting an introduction to the space environment with an emphasis on those facets of the environment that may degrade spacecraft subsystems. The objective is to obtain an understanding of the relationship between the space environment and spacecraft, or space instrument, operating principles and design alternatives. This text is formatted to illustrate these relationships by presenting a description of the space environment, a discussion of the ways in which the environment may interact with an orbiting spacecraft, and finally by relating the various environmental interactions to spacecraft design specifics. Understanding these relationships is important not only to spacecraft designers, who must develop a spacecraft capable of operating in a specified orbital environment, but also to payload providers, who must provide instrumentation capable of delivering high-quality data under potentially adverse conditions.

1.2 The Space Environment

The first problem encountered in the study of space environment effects is that of defining the various environments. Historically, these definitions have varied depending on the nature of the text or the objective of the author.[2-4] In order to reflect the fundamental differences in the nature of the local space environment, this text will group space environment effects into five categories: vacuum, neutral, plasma, radiation, and micrometeoroid/ orbital debris (MMOD).

At an altitude of 300 km, a typical operational altitude for the space shuttle, the atmospheric mass density is approximately ten orders of magnitude less than that found at sea level. The vacuum environment therefore describes those phenomena that are related to operation under vacuumlike conditions.

A ten order of magnitude decrease in mass density is significant in relative terms, but in absolute terms the reduction in atmospheric density is from ~ 10^{27} m^{-3} to ~ 10^{17} m^{-3}. These neutral particles can interact with a spacecraft both mechanically, through the kinetic energy of impact, or chemically, through the reactive nature of the neutrals themselves. The neutral environment therefore describes those phenomena that are related to the presence of these neutral constituents.

At 300 km about 1% of the neutral atoms are ionized, stripped of an electron, by the solar ultraviolet radiation. These charged particles form the plasma environment, which may give rise to a completely different category of interactions. These particles may charge the exterior surfaces of a vehicle to large voltages. As a result, data collected by scientific instrumentation may be biased, or arcing on the spacecraft may ensue. Consequently, the plasma environment includes the phenomena associated with charged particles having energy in the range KeV or less.

Some charged particles will be encountered with energies in the range MeV or higher. These particles will exert a completely different influence on a spacecraft than their lower energy counterparts. They will penetrate through the bulk of many materials, altering their physical structure. These higher-energy particles constitute the fourth category, the radiation environment.

Finally, a spacecraft may also encounter micron-sized (or larger) pieces of naturally occurring dust or manmade debris that composes the micrometeoroid/orbital debris (MMOD) environment. As we will see, at orbital velocities these particles will give rise to surface erosion and have the potential to terminate a spacecraft's useful life. The interactions associated with each of these environments are summarized in table 1.2.

Depending on the spacecraft orbit, the magnitude of these interactions may range from negligible to mission threatening. Chapters 2–6 are devoted to an in-depth discussion of each of the various environments and their specific effect on spacecraft. In each case, emphasis is placed on understanding the basic nature of the environment and of the potential interactions between the environment and spacecraft engineering subsystems. Chapter 7 summarizes the main conclusions and examines the possibility for synergistic effects between the various environments. Before beginning an in-depth discussion, however, it is appropriate to review some of the fundamentals of spacecraft design, the Earth's magnetic field, and the solar-planetary relationship. The results of these introductory sections will be referred to in later chapters as appropriate.

Table 1.2
Space Environment Effects

Environment	Effect
Vacuum	*Solar UV Degradation* *Contamination*
Neutral	*Mechanical Effects* Aerodynamic drag Physical sputtering *Chemical Effects* Atomic oxygen attack Spacecraft glow
Plasma	*Spacecraft Charging* Shift in ground potential *Electrostatic Discharging* Dielectric breakdown Gaseous arc discharge *Enhanced sputtering* *Reattraction of contamination*
Radiation	*Total Dose Effects* Solar cell degradation Sensor degradation Electronics degradation *Single Event Effects* Upsets, Latchup, . . .
Micrometeoroid/ Orbital Debris	*Hypervelocity Impacts*

1.3 Spacecraft Design

Regardless of the specific nature of the payload, all spacecraft must perform certain basic functions in order to enable the payload to function properly. The spacecraft must have a propulsion system to boost the spacecraft to its intended orbit and to provide the spacecraft with a means of maintaining the orbit, or de-orbiting at the end of the mission. There must be an Electrical Power System (EPS) to provide power to the payload and other subsystems. A Thermal Control System (TCS) must maintain the spacecraft within the proper operating temperature range. An Attitude Determination and Control (ADC) system is needed to orient the vehicle and to point the payload at its desired point of reference. A variety of avionics is needed to route electrical commands around the vehicle; a Telemetry, Tracking and

Communications (TT&C) system is needed to transmit data to the ground and receive instructions from the Earth; and finally, a physical structure is needed to accommodate the systems and payload. Typically, subsystem design engineers have a variety of alternatives to examine in performing trades to identify the optimal systems level solution (table 1.3).[2] In general, the most critical components from the point of view of space environment effects are those components that are directly exposed to the space environment. The specifics of the various subsystem design alternatives will be explored in more detail in future chapters as needed. For more specific information on spacecraft vehicle and mission design the reader is referred to the works by Griffin and French and Brown.[5,6]

Table 1.3
Spacecraft Engineering Subsystems

Subsystem	Purpose	Key Features
Attitude determination & control	Vehicle stability & pointing control	Reaction heels Momentum wheels Sun/Earth sensors Magnetic torquers
Avionics	Data and command relay to payloads and subsystems	Data bus Processing Memory
Electrical power	Power generation and distribution	Solar arrays Batteries Load control electronics
Propulsion	Maneuver vehicle into desired orbit	Thrusters Fuel Tanks, plumbing
Structures	Integrity during launch and maneuvers	Bulkheads Mechanisms
Telemetry, tracking & communications	Command and data handling with the ground	Transmitters Receivers Antennas
Thermal control	Maintain temperature balance	Radiators, heaters Heat pipes Multilayer insulation Anodized Aluminum

Most spacecraft can be grouped into one of three orbital altitude ranges as shown in figure 1.1. Orbits with a perigee of less than about 1000 km are said to be in *low Earth orbit* (LEO). This is the region frequented by the space shuttle and is typically reserved for the largest operational payloads (e.g., the space station, *Skylab*, *Mir*, *Salyut*) or spacecraft that need a close view of the Earth (e.g., LANDSAT, TIROS, DMSP). The 1000–2000 km altitude range is referred to as *mid-Earth orbit* (MEO), or sometimes *high Earth orbit* (HEO), and is frequented by reconnaissance satellites placed in highly elliptical orbits. Finally, the geosynchronous (GEO) altitude, 35,800 km, is popular with various surveillance and communications spacecraft. A spacecraft's orbital altitude will have a major impact on the type and magnitude of space environment effects experienced by a spacecraft.

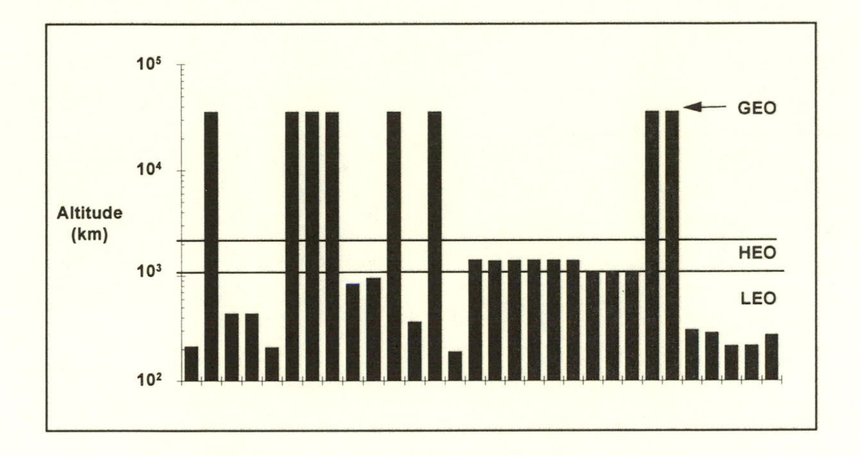

Fig. 1.1 Perigee of all spacecraft launched in fourth quarter 1991.

A second important factor is the orbital inclination, the angle between the Earth's equator and the orbital plane. For example, an orbital inclination of about 99° (the actual value is a function of altitude) places the spacecraft in an orbital plane that precesses at the same rate as the Earth orbits around the Sun. This provides the spacecraft with a constant Sun angle and is referred to as a *Sun-synchronous orbit*. Different inclination orbits are used for a variety of reasons; however, a primary driver of orbital inclination is the latitude of the launch site. A given launch vehicle can launch the heaviest possible payload into an inclination equal to the latitude of the launch site. The inclination of numerous launch facilities is presented in table 1.4. Differences in environmental interactions that are dependent on altitude and/or inclination will be emphasized where appropriate in the following chapters.

Table 1.4
Worldwide Launch Facility Latitudes

Facility	Latitude	Facility	Latitude
Australia		*Japan*	
Woomera	31° 07' S	Kagoshima	31° 14' N
		Osaki	30° 24' N
China		Takesaki	30° 23' N
East Wind	40° 25' N		
Tai-yuan	37° 46' N	*India*	
Wuzhai	38° 35' N	Thumba	08° 35' N
Xichang	28° 06' N		
		Israel	
Commonwealth of Independent States		Yavne	31° 31' N
Plesetsk	62° 48' N	*United States*	
Kapustin Yar	48° 24' N	Eastern range	28° 30' N
Tyuratam	45° 54' N	Wallops	37° 51' N
		Western range	34° 36' N
Europe (ESA)			
Kourou	05° 32' N		
San Marco (Italy)	02° 56' S		

1.4 The Earth's Magnetic Field

As we will see in the chapters on the plasma and radiation environments, the Earth's magnetic field plays an important part in defining the near-Earth environment. The Earth's magnetic field is essentially a dipole field that is complicated by the fact that (1) the magnetic axis is not aligned with the geographical axis, and (2) the Sun's magnetic field causes perturbations at altitudes higher than about 2000 km.

The north magnetic pole is approximately 11.5° south of the north geographic pole at 78.3° N, 69° W, near Thule, Greenland. The south magnetic pole is at 78.3° S, –111° E, near Vostok Station, Antarctica. The magnetic potential of a dipole, V, is given by

$$V = \frac{\vec{M} \cdot \vec{r}}{r^3}, \tag{1.1}$$

where r (m) is the distance from the center of the field and M (Tesla m^3) is the magnetic moment. For the Earth, M is approximately 8×10^{14} Tesla m^3 or 8×10^{24} gauss cm^3. In spherical coordinates, with θ measured from the dipole axis (geomagnetic colatitude), the dipole field is given by $B = \nabla V$ or

$$
\begin{aligned}
B_r &= \frac{\partial V}{\partial r} = -\frac{2M}{r^3}\cos\theta \\
B_\theta &= \frac{1}{r}\frac{\partial V}{\partial \theta} = -\frac{M}{r^3}\sin\theta \\
B_\phi &= \frac{1}{r\sin\theta}\frac{\partial V}{\partial \phi} = 0.
\end{aligned}
\tag{1.2}
$$

The total intensity of the field is given by

$$B = \frac{M}{r^3}\left[3\cos^2\theta + 1\right]^{1/2}. \tag{1.3}$$

(Note that some authors define the field in terms of the dipole latitude, which would be measured from the geomagnetic equator toward the north geomagnetic pole, opposite from the convention used here.) The resulting field strength found at the surface of the Earth is approximately 30 microTesla (0.3 gauss) at the equator and 60 microTesla (0.6 gauss) at the poles. As will be shown in chapter 3, in the absence of electric fields charged particles are constrained to gyrate around the magnetic field lines. Consequently, the bulk motion of charged particles can only be along the magnetic field lines. At any given point, the slope of the magnetic field line is given by

$$\frac{B_\theta}{B_r} = \frac{r\partial\theta}{\partial r} = \frac{1}{2}\tan\theta. \tag{1.4}$$

Integrating this equation gives the constraint

$$r = LR_E\sin^2\theta, \tag{1.5}$$

where R_E is defined to be the Earth radius and L is a constant of integration. Charged particles in the presence of a dipole field are constrained to gyrate around the magnetic field line described by the constant L. Because of this, one set of coordinates encountered in studies involving the Earth's magnetic field is the B-L coordinate system, which is illustrated in figure 1.2.

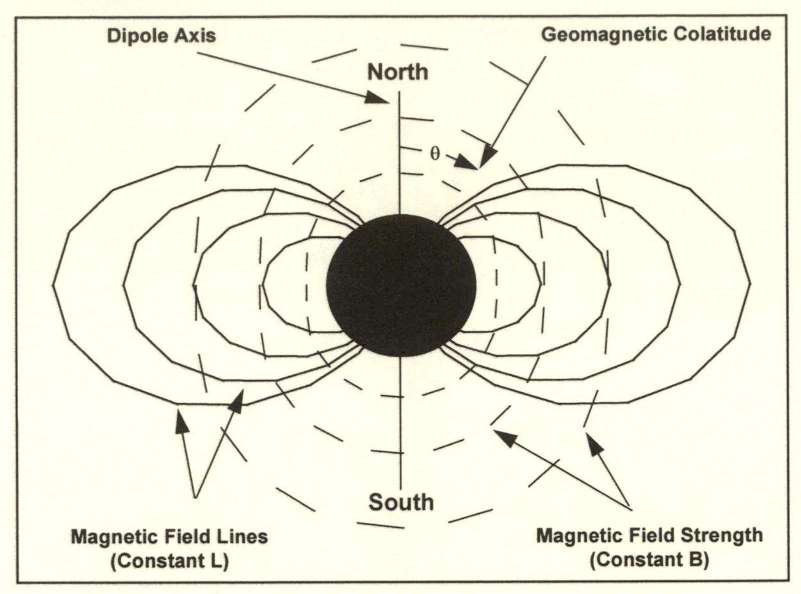

Fig. 1.2 The magnetic field lines and magnetic field strength.

Any field derivable from a potential function may be expressed in terms of a multipole expansion of the potential. That is,

$$V = a\sum_{n=1}^{\infty}\sum_{m=1}^{\infty} P_n^m(\cos\theta) \times \tag{1.6}$$

$$\left[\left(\frac{a}{r}\right)^{n+1}\left(g_n^m \cos m\phi + h_n^m \sin m\phi\right) + \left(\frac{a}{r}\right)^{-n}\left(A_n^m \cos m\phi + B_n^m \sin m\phi\right)\right]$$

where r, θ, ϕ are the geographical polar coordinates of radial distance, colatitude, and east longitude, a is the radius of the Earth (R_E), the functions P_n^m (cosθ) are the *Legendre polynomials*, also known as the *Schmidt functions*, and the coefficients g_n^m, h_n^m, A_n^m, and B_n^m are referred to as the *Schmidt coefficients*. The Schmidt coefficients are widely used in quoting the results of analyses or predictions. Physically, the terms g_n^m and h_n^m refer to field sources internal to the Earth while the terms A_n^m and B_n^m arise from external currents in the outer atmosphere. Many models of the Earth's magnetic field, such as the International Geomagnetic Reference Field (IGRF), specify the Earth's magnetic potential in terms of Legendre polynomials.

1.5 The Solar-Planetary Relationship

Virtually every energy conversion process on the Earth can eventually be related back to the fact that the Sun is shining. The Sun is the largest mass in the solar system and is the predominant source of energy for the Earth. On average, the Sun deposits 1371 ± 5 W/m^2 of energy at the top of the Earth's atmosphere. This varies from a high of 1423 W/m^2 at Sun-Earth perigee to a low of 1321 W/m^2 at Sun-Earth apogee. The Sun is composed primarily of hydrogen and generates its energy through the fusion of hydrogen into helium. The primary fusion reaction is

$$ {}_1^2H + {}_1^2H \rightarrow {}_2^4He + {}_1^0n. \tag{1.7}$$

The Sun is characterized by a variety of phenomena and is divided into various regions depending on the nature of the physical processes found there (fig. 1.3). The majority of the solar energy reaching the Earth originates in the outermost layer of the Sun called the corona. Although the interior of the Sun is quite hot, ~ 15 million K, the exterior is, in comparison, relatively cool, and the spectrum of the solar output resembles that of a blackbody at 5760 (fig. 1.4). This may be compared with the electromagnetic spectrum in figure 1.5.

1.5.1 The 11-Year Solar Cycle

As has been documented for centuries, the output of the Sun is not constant, but varies slightly over an 11-year cycle.[7] Historically, this variation was noticeable by observing the number of sunspots present (table 1.5 and fig. 1.6). In modern times the total integrated solar output is monitored. The variation in solar output from one solar cycle to the next is quite small, less than 0.1%; however, the variations are significant for a number of reasons. As we will see in chapter 2, the Earth's atmosphere absorbs all ultraviolet (UV) radiation less than 0.3 microns in wavelength. As UV output changes over the solar cycle, the amount of energy absorbed by the Earth's atmosphere changes and the atmosphere responds accordingly. Heating a gas will cause the gas to expand. Consequently, the density, and temperature, of the neutral atmosphere will change in response to the solar cycle. This in turn will affect plasma density at the lower altitudes.

The 11-year solar cycle is monitored by tracking sunspot activity, which is defined as

$$R = k(10g + s), \tag{1.8}$$

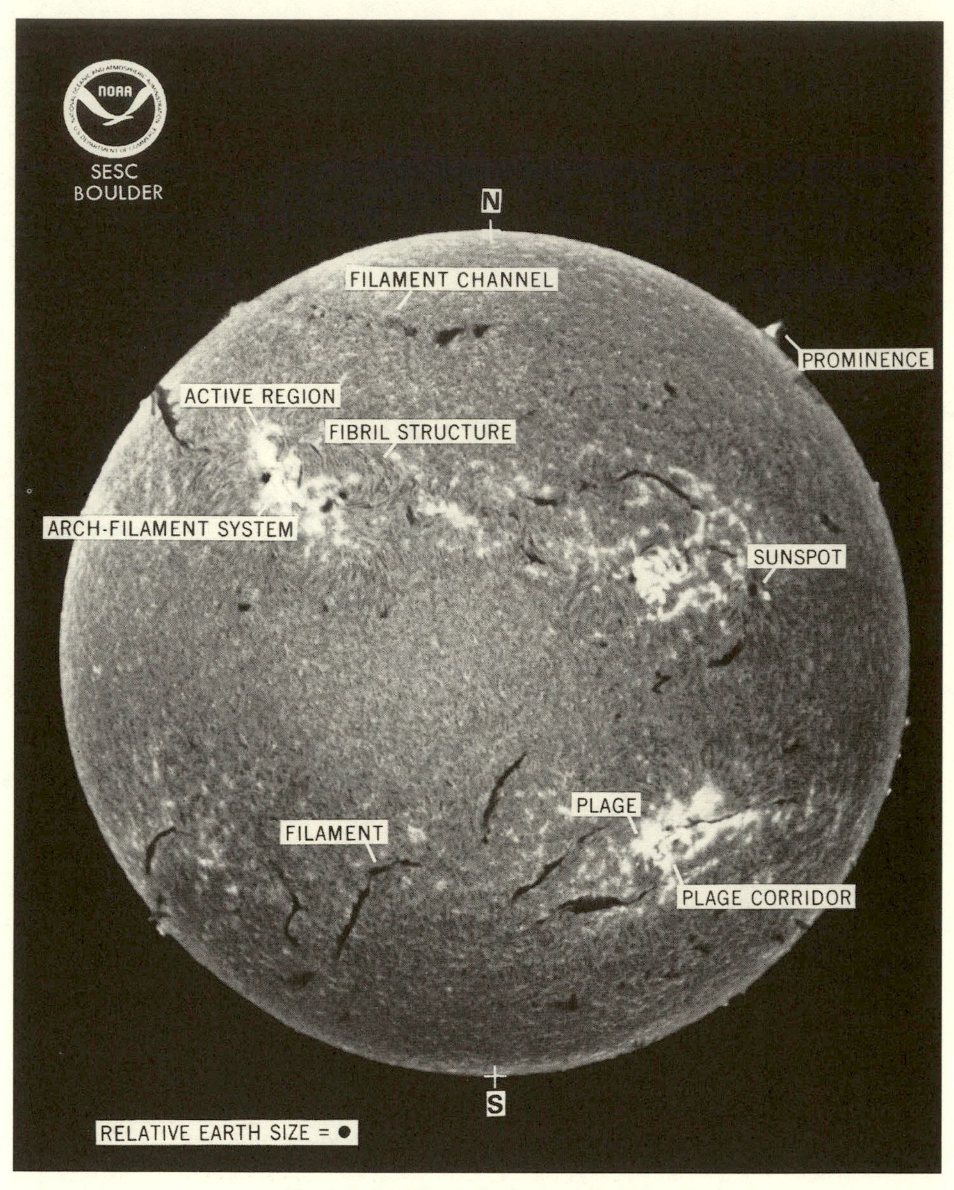

Fig. 1.3 The Sun in hydrogen alpha light.
(Photograph courtesy of NOAA)

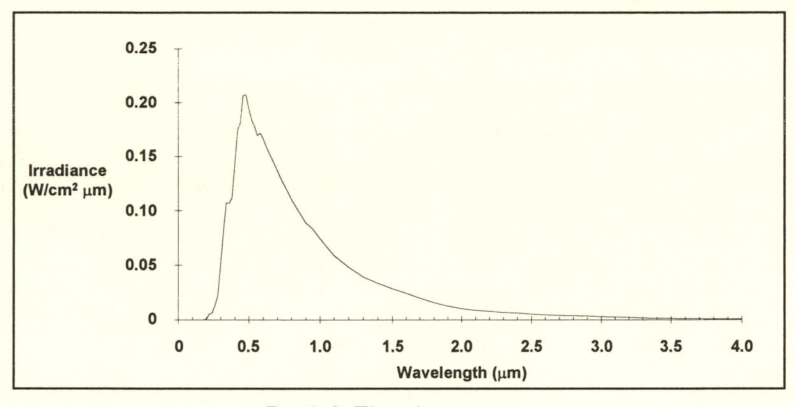

Fig. 1.4 The solar spectrum.

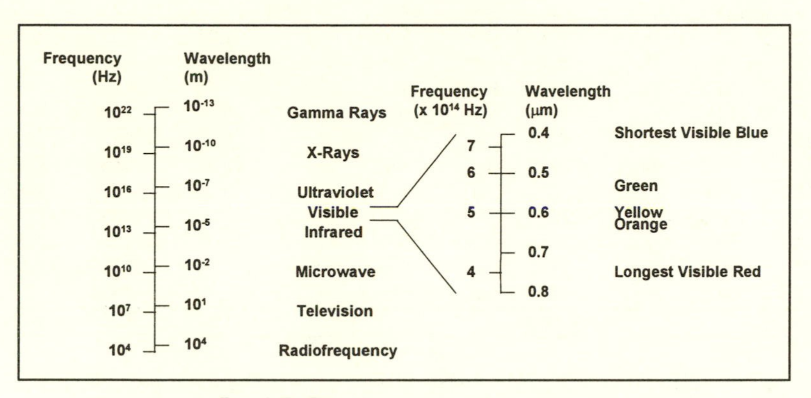

Fig. 1.5 The electromagnetic spectrum.

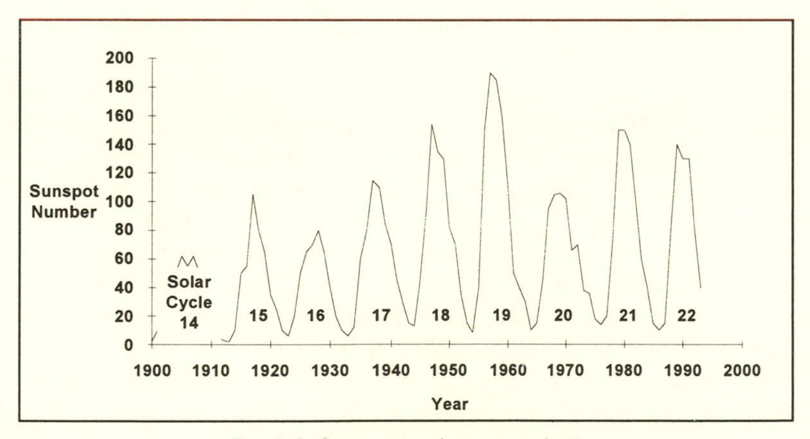

Fig. 1.6 Sunspot number versus time.

Table 1.5
Solar Cycle Activity

	Start		Solar Max.		End		Length	Rise - Max	Max - End
No.	Year	Mo.	Year	Mo.	Year	Mo.	Year	Year	Year
1	1755	Mar.	1761	Jun.	1766	May	11.25	6.25	5.00
2	1766	Jun.	1769	Sep.	1775	May	9.00	3.25	5.75
3	1775	Jun.	1778	May	1784	Aug.	9.25	2.92	6.33
4	1784	Sep.	1788	Feb.	1798	Apr.	13.67	3.42	10.25
5	1798	May	1805	Feb.	1810	Jul.	12.25	6.75	5.50
6	1810	Aug.	1816	Apr.	1823	Apr.	12.75	5.67	7.08
7	1823	May	1829	Nov.	1833	Oct.	10.50	6.50	4.00
8	1833	Nov.	1837	Mar.	1843	Jun.	9.67	3.33	6.33
9	1843	Jul.	1848	Feb.	1855	Nov.	12.42	4.58	7.83
10	1855	Dec.	1860	Feb.	1867	Feb.	11.25	4.17	7.08
11	1867	Mar.	1870	Aug.	1878	Nov.	11.75	3.42	8.33
12	1878	Dec.	1883	Dec.	1890	Feb.	11.25	5.00	6.25
13	1890	Mar.	1894	Jan.	1901	Dec.	11.83	3.83	8.00
14	1902	Jan.	1906	Feb.	1913	Jul.	11.58	4.00	7.58
15	1913	Aug.	1917	Aug.	1923	Jul.	10.00	4.00	6.00
16	1923	Aug.	1928	Apr.	1933	Aug.	10.08	4.67	5.42
17	1933	Sep.	1937	Apr.	1944	Jan.	10.42	3.58	6.83
18	1944	Feb.	1947	May	1954	Mar.	10.17	3.25	6.92
19	1954	Apr.	1958	Mar.	1964	Sep.	10.5	3.92	6.58
20	1964	Oct.	1968	Nov.	1976	May.	11.67	4.08	7.58
21	1976	Jun.	1979	Dec.	1986	Aug.	10.25	3.50	6.75
22	1986	Sep.	1989	Jul.				2.83	
			Averages				**11.02**	**4.22**	**6.73**

where s = number of individual spots, g = number of sunspot groups, and k is an observatory factor (equal to 1 for the Zurich observatory and adjusted for all other observatories to obtain approximately the same R number). Equivalently, the variation can be monitored by tracking the solar output in the UV region of the spectrum, which must be done from space. Variations here can be as high as a factor of 10 or more. Consequently, an equally accepted means of measuring the intensity of the solar output is the solar flux at 10.7 cm wavelength. This is often reported as the F10.7 value, which is measured in units of Jansky = 10^{-26} W/m^2 Hz, or equivalently in solar flux units (sfu) = 10^4 Jansky. Each 11-year solar cycle is numbered sequentially, with the first solar cycle corresponding to observations made in the 1750s. Solar cycle 22 will end in the summer of 1997, at which point solar cycle 23 will begin. Times of low solar activity, 1997, are known as *solar minimum*, while times of high solar activity, 2002, are known as *solar maximum*.

Once a solar cycle is underway it is possible to predict the smoothed monthly average F10.7 values, based on linear regression analysis of past cycles, for the remainder of the cycle (fig. 1.7). However, due to the large uncertainties associated with past observations, predictions of future solar cycles are not as statistically accurate. NASA provides a best estimate (50%), an upper bound (97.7%), and a lower bound (2.3%) in its predictions for future solar cycles (fig. 1.8). As can be inferred from figure 1.8, about one year after the start of a solar cycle it is possible to predict more accurately whether the cycle will be large, average, or small.

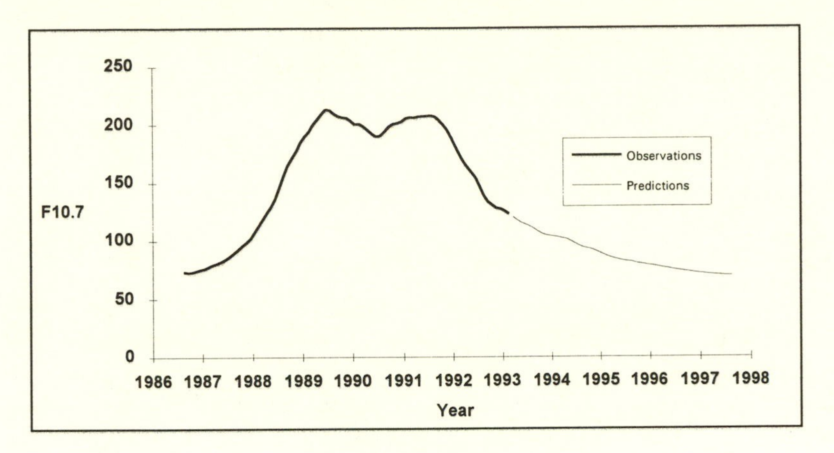

Fig. 1.7 F10.7 values for solar cycle 22.

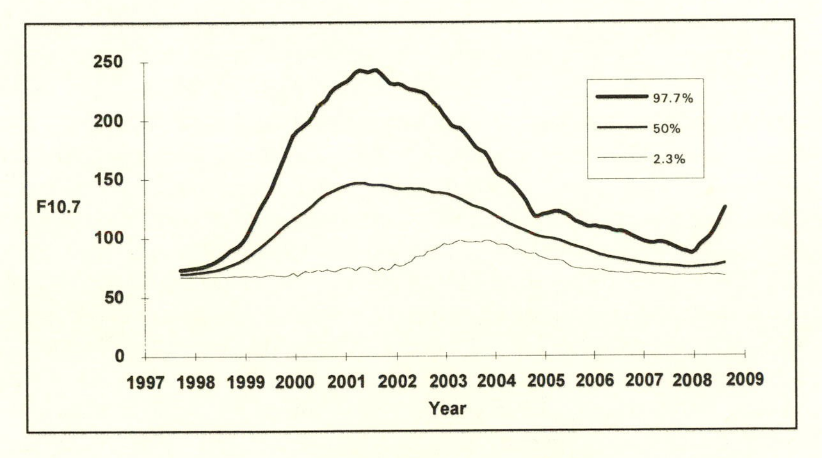

Fig. 1.8 F10.7 values for solar cycle 23.

Note that the statistical probabilities 97.7%, 50%, and 2.3% represent the probability that the prediction will overestimate the magnitude of the solar cycle. That is, the 97.7% curve implies that there is a 97.7% probability that the next cycle will be less intense than the curve indicates. The value of F10.7 used in calculations is therefore related to the margin desired in the results. As shown in table 1.6, an average conservative value of F10.7 under nominal conditions is 100.

Table 1.6
F10.7 Values for Various Solar Cycle Conditions

	F10.7		
Confidence (%)	**Solar Min.**	**Average**	**Solar Max.**
2.3	75	85	95
50.0	75	100	150
97.7	75	150	250

1.5.2 The Solar Wind and Solar Flares

In addition to the electromagnetic radiation produced by the Sun, a variety of processes in the Sun also give rise to a small amount of corpuscular radiation called the *solar wind.* The solar wind is composed mainly of protons, which have a directed velocity on the order of 375 km/s and a density of ~ 5 cm^{-3} at 1 astronomical unit (AU), the distance from the Sun to the Earth. The charged particles in the solar wind follow trajectories tied to the Sun's magnetic field lines, which have an intensity of about 5 nanoTesla at 1 AU. The Sun is observed to rotate on its axis with a period of rotation of 25 days at the equator, with the higher latitudes requiring a few days longer. As a result of the rotation, the trajectory of particles in the solar wind resembles the flow of water from a rotating sprinkler head (fig. 1.9). The solar wind is occasionally supplemented by additional bursts of radiation associated with visual events on the Sun called *solar flares.* A flare is a sudden brightening of the chromosphere and may be accompanied by outbursts of energetic particles. There are five classifications of solar flares (table 1.7), with each classification being further designated with the letter F, N, or B to denote faint, normal, or bright. Solar flares may occur at either optical or X ray wavelengths, with X ray flares being of greater concern to spacecraft (table 1.8). Solar activity, as measured by enhanced X ray emission, is monitored by the Space Environment Services Center in Boulder, Colorado. As shown in table 1.9, five standard terms are used to describe the solar activity observed or expected within a 24-hour period. The SESC issues alerts and warnings of a variety of solar-geophysical phenomena that may directly or indirectly affect spacecraft performance. Flares may also be accompanied by a variety of radio emissions (table 1.10). These radio bursts may temporarily interrupt communications with a spacecraft or degrade its effectiveness.

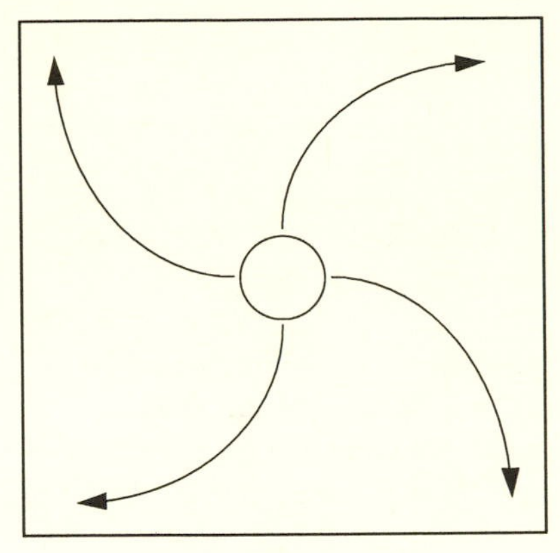

Fig. 1.9 The solar wind.

Table 1.7
Solar Flare Class

Importance	Square Degrees
0	≤ 2.0
1	2.1–5.1
2	5.2–12.4
3	12.5–24.7
4	≥ 24.8

1 square degree = $(1.214 \times 10^4 \text{ km})^2$
= 48.5×10^{-6} of the visible hemisphere.

Table 1.8
X ray Flare Class

Classification	(W m^{-2})	(ergs cm^{-2} s^{-1})
B	$I < 10^{-6}$	$I < 10^{-3}$
C	$10^{-6} \leq I < 10^{-5}$	$10^{-3} \leq I < 10^{-2}$
M	$10^{-5} \leq I < 10^{-4}$	$10^{-2} \leq I < 10^{-1}$
X	$I \geq 10^{-4}$	$I \geq 10^{-1}$

Note: The peak burst intensity is measured in the 0.1 to 0.8 nm band.

Table 1.9
Solar Activity by X ray Flare Class

Very low	Low	Moderate	High	Very high
Events less than C-class	C-class Events	Isolated (1–4) M-class Events	Several (>4) M-class Events, or Isolated (1–4) M5-class Events	Several (>4) M5-class Events

Of greater concern for the long-term survivability of a spacecraft are energetic fluxes of particles, mainly protons, associated with some flares. These particle fluxes, referred to as *coronal mass ejections* (CMEs), are often, but not always, associated with erupting prominences and flares. CMEs may be responsible for generating a significant number of solar protons that may turn out to be the most important factor in contributing to a spacecraft's total radiation dose. This will be discussed further in chapter 5.

Table 1.10
Solar Flare Radio Burst Type

Type	Description
I	Storm composed of many short, narrow-band bursts in the range 300 Hz to 50 MHz of extremely variable intensity. Storm may last from several hours to several days.
II	Narrow-band emission (sweep) that begins in the range 300 MHz and sweeps slowly (tens of minutes) toward 10 MHz. Emissions occur in loose association with major flares and are indicative of shock waves moving through the solar atmosphere.
III	Narrow-band bursts that sweep rapidly (seconds) from 500 Hz to 0.5 MHz. Bursts often occur in groups and are an occasional feature of complex solar active regions.
IV	A smooth continuum of broad-band bursts primarily in the range 300 Hz to 30 MHz. Bursts occur with some major flare events: they begin within 2 minutes after flare maximum and can last for hours.
V	Short duration (a few minutes) continuum noise in the 10 MHz range, usually associated with type III bursts.

1.5.3 Geomagnetic Storms

The boundary between the region where the Sun's magnetic field dominates and the region where the Earth's magnetic field dominates is called the *magnetopause* (fig. 1.10). The Sun's magnetic field fluctuates in response to various solar phenomena, on time scales as short as seconds. Consequently, the magnetopause moves in response to the Sun's field. Normally, in the sunward direction the magnetopause is at about 10 Earth radii, but it occasionally gets pushed to within the geosynchronous orbit (6.6 Earth radii). These magnetic fluctuations, or magnetic storms, are typically quite small, being on the order of nanoTeslas. However, as a result of the storms higher-energy, charged particles may impact spacecraft in GEO, which may increase the radiation dose to the vehicle and/or result in charging of surfaces to high values. Effects may also be noticed on the Earth's surface. On 13 March 1989 over 6 million people in Quebec, Canada, were without electrical power for nine hours as the result of an equipment failure attributed to an induced current caused by a geomagnetic storm.

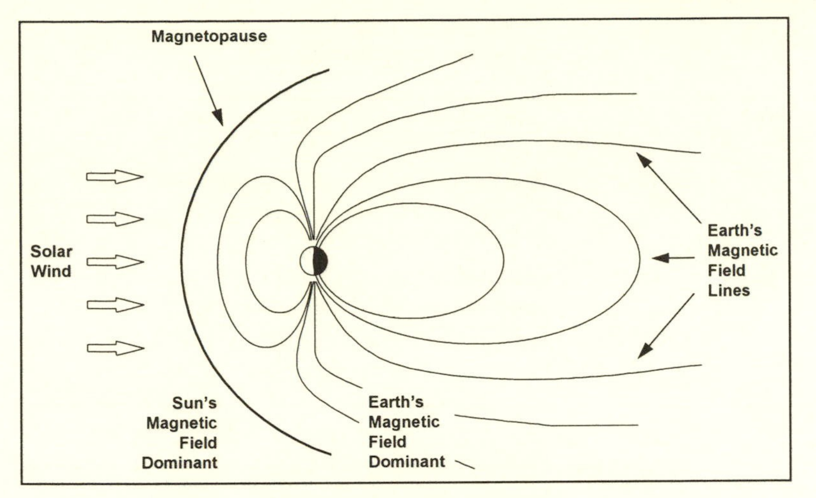

Fig. 1.10 The magnetopause.
(Reprinted with permission of VCH Publishers, ©1991.)

Because of the interaction between the solar magnetic field and the Earth's magnetic field, the magnetic field strength encountered at higher orbital altitudes is also seen to vary with an 11-year cycle. A variety of parameters is used to describe fluctuations in the strength of the geomagnetic field. The K index, a quasi-logarithmic value ranging from 0 (calm) to 9 (greatly disturbed), is a measure of the general level of magnetic activity caused by the solar wind. Values of K are made at a number of worldwide observation stations and are used to compute the local station index, K_s, which is a modified version of K accounting for local and seasonal magnetic variations. The K_s indices are used to compute a planetary index, K_p. Because K, K_s, and K_p are quasi-logarithmic they are difficult to incorporate into numerical models. Consequently, the K indices are converted to roughly linear a_k indices, as shown in table 1.11. Daily averages of a_k and a_p are in turn used to compute the daily station index, A_k, and the average planetary index (also called the geomagnetic index), A_p (table 1.12). Past observations and future predictions of A_p are illustrated in figures 1.11 and 1.12, respectively. The relation between the various space environment effects and the solar cycle will be discussed in more detail in the following chapters.

Table 1.11
The a_k Index

K	0	1	2	3	4	5	6	7	8	9
a_k	0	3	7	15	27	48	80	140	240	400

Table 1.12
Geomagnetic Activity

Category	a_k index
Quiet	0–7
Unsettled	8–15
Active	16–29
Minor storm	30–49
Major storm	50–99
Severe storm	100–400

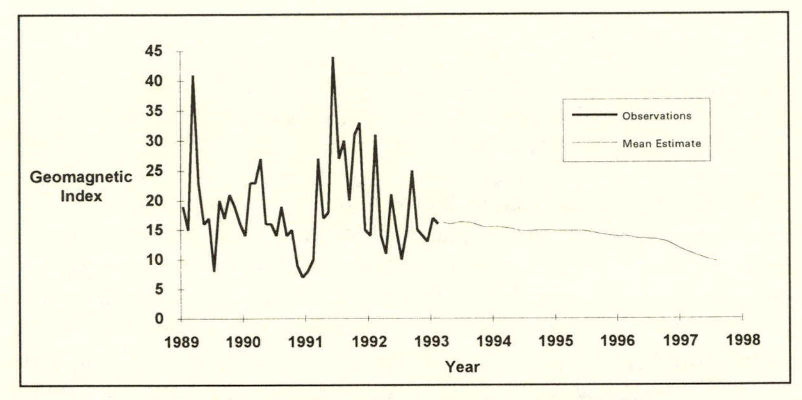

Fig. 1.11 The geomagnetic index for solar cycle 22.

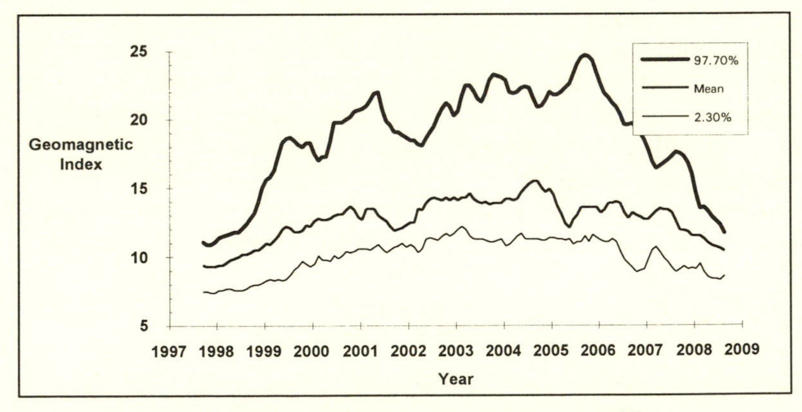

Fig. 1.12 The geomagnetic index for solar cycle 23.

1.6 Exercises

1. From the Universal Law of Gravitation,

$$F = G\frac{M_E m_{s/c}}{(R_E + h)^2},$$

where G (N m^2/kg^2) is the Universal Gravitational Constant, M_E (kg) is the mass of the Earth, $m_{s/c}$ (kg) is the mass of the spacecraft, R_E (m) is the radius of the Earth, and h (m) is the spacecraft orbital altitude, and from the relation between acceleration, velocity, and radius for an object undergoing circular motion

$$a = \frac{v^2}{(R_E + h)},$$

a.) derive expressions for the orbital velocity and orbital period as a function of altitude; and

b.) graph velocity and period vs. altitude.

1.7 Applicable Standards

ASTM E 490, *Standard Solar Constant and Air Mass Zero Solar Spectral Irradiance Tables*, 27 September 1973.

1.8 References

1. Anderson, B. J., ed., *Natural Orbital Environment Guidelines for Use in Aerospace Vehicle Development*, NASA Technical Manual 4527, June 1994.
2. Tascione, T., *Introduction to the Space Environment* (Malibar, FL: Orbit Book Company, 1988).
3. Damon, T. D., *Introduction to Space—The Science of Spaceflight* (Malibar, FL: Orbit Book Company, 1989).
4. Jursa, A. S., ed., *Handbook of Geophysics and the Space Environment* (Hanscom Air Force Base, MD: Air Force Geophysics Laboratory, 1985).

5. Griffin, M. D., and French, J. R., *Space Vehicle Design* (Washington, DC: American Institute of Aeronautics and Astronautics, 1991).
6. Brown, C. D., *Spacecraft Mission Design*, (Washington, DC: American Institute of Aeronautics and Astronautics, 1992).
7. Schove, D. J., ed., *Sunspot Cycles*, vol. 68, Benchmark Papers in Geology (Stroudsburg, PA: Hutchinson Ross Publishing Company, 1983).

2 The Vacuum Environment

There's nothing out there. That's why it's called space.
—Carl Sagan, TV Series Cosmos

2.1 Overview

At 100 km altitude the ambient atmospheric pressure is over six orders of magnitude less than that found at sea level. Even before a spacecraft is released from the launch vehicle, the surrounding pressure will drop to vacuumlike levels justifying the assertion that, relative to the Earth's surface, space is a vacuum. Consequently, it is appropriate first to discuss the vacuum environment and its effect on spacecraft design. Designing a spacecraft to operate under vacuumlike conditions places constraints on choices of materials and thermal control. Upon exposure to very low pressures, many materials will exhibit a mass loss through a process called *outgassing*. Essentially, volatile materials may escape the attraction to the surface and be released into the surrounding atmosphere. This is the source of the "new car smell" associated with a motor vehicle. As is often experienced, the new car smell may be accompanied by the buildup of a thin contaminant film on the inside surfaces of the vehicle. In space, outgassing materials may deposit contaminants onto sensitive surfaces such as thermal control panels, solar arrays, or optics, thereby altering their thermal or optical properties. When the nearest bottle of window cleaner is 350 km away, and moving at a relative speed of 8 km/s, maintaining surface cleanliness is mandatory for mission success. An additional challenge to operating in a vacuum is thermal control. Without the presence of an atmosphere to aid in convective heat transfer, a spacecraft may cool itself only by conduction or radiation. Conduction may be relied upon to transfer heat between various parts of a spacecraft, but radiation is the primary heat transfer mechanism between a spacecraft and its surrounding environment. An object will absorb heat, Q (W), from the Sun according to the relation

$$Q_{in} = \alpha_s A_n S, \tag{2.1}$$

where α_s is the material's solar absorptance, A_n (m^2) is the surface area normal to the solar flux, and S (W/m^2) is the solar flux per unit area at the spacecraft orbit. As discussed in the preceding chapter, the average value for S, above the Earth's atmosphere, is 1371 ± 5 W/m^2 at 1 AU. An object will also radiate heat to its surrounding environment according to the relation

$$Q_{out} = \varepsilon A_{tot} \sigma T^4, \tag{2.2}$$

where ε is the material emittance, A_{tot} (m^2) is the total surface area, T (K) is the object temperature, and σ (W/m^2K^4) is Boltzmann's constant. (Numerical values for all pertinent physical constants are provided in appendix 3.) In a first approximation one can assume that the material temperature is much greater than that of the surrounding space environment so that radiation to the vehicle from sources other than the Sun is small in comparison. (Note that the Earth albedo may be significant in LEO, however.) Subject to this constraint, the equilibrium temperature of the material is approximated by

$$T_{s/c} = \left(\frac{\alpha_s}{\varepsilon}\right)^{1/4} \left(\frac{SA_n}{\sigma A_{tot}}\right)^{1/4}. \tag{2.3}$$

As an example, the blackbody ($\alpha_s/\varepsilon = 1$) temperature of a sphere at the distance of the orbit of the planets is shown in figure 2.1. Like living matter, machinery works best at about room temperature and thermal control is an important aspect of spacecraft design.

All excess heat generated by a spacecraft must be dissipated by some means, most often through the use of thermal radiators. To minimize spacecraft mass and volume, materials having low values of α_s and high values of ε are used for radiator surfaces. Thermal balance can be maintained over the spacecraft's lifetime only if the radiator maintains its thermal properties, that is, its α_s/ε ratio. Solar absorptance for a given material is defined by integrating over the thermal spectrum,

$$\alpha_s = \frac{\int \alpha_s(\lambda) S(\lambda) d\lambda}{\int S(\lambda) d\lambda}, \tag{2.4}$$

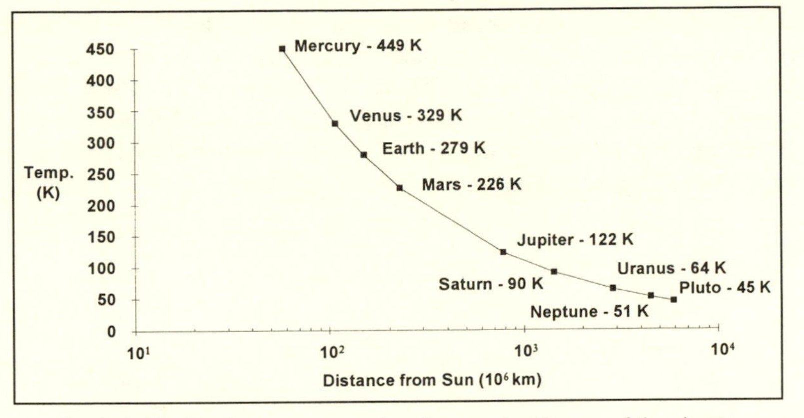

Fig. 2.1 Blackbody temperature of a sphere at the distance of the planets.

where $\alpha_s(\lambda)$ is the material absorptance as a function of wavelength, $S(\lambda)$is the solar output as a function of wavelength, and λ is the wavelength. Note that $\alpha_s(\lambda)$ must be experimentally determined and $S(\lambda)$ is given in figure 1.4. Typical values of α_s and ε are listed in table 2.1 for common spacecraft materials.[1]

Table 2.1
Absorptance/Emittance of Typical Spacecraft Materials

Material	α_s	ε	α_s/ε
Aluminum	0.10	0.05	2.000
Black Paint	0.97	0.89	1.090
FEP—2 mil	0.06	0.68	0.088
FEP—5 mil	0.11	0.80	0.138
Fused Silica	0.08	0.81	0.099
Gold	0.21	0.03	7.000
Grafoil	0.66	0.34	1.941
Kapton	0.48	0.81	0.593
Kapton, coated	0.40	0.71	0.563
Kapton, Au film	0.53	0.42	1.262
OSR, coated	0.09	0.76	0.118

2.2 Vacuum Environment Effects

As shown in figure 2.2, on orbit most spacecraft witness a degradation in thermal control associated with changes in radiator absorptance.[2-5] A change in α_s of 0.2 would be catastrophic for a radiator designed with a beginning of life value of 0.08. Solar absorptance can be degraded by a number of environmental factors such as radiation, micrometeoroid/orbital debris impact, arcing, sputtering, contamination, or the solar UV. These last two topics are vacuum environment effects and are discussed in the following sections. As we will see, changes in α_s can be greatly reduced through the proper choice of materials and spacecraft configuration.

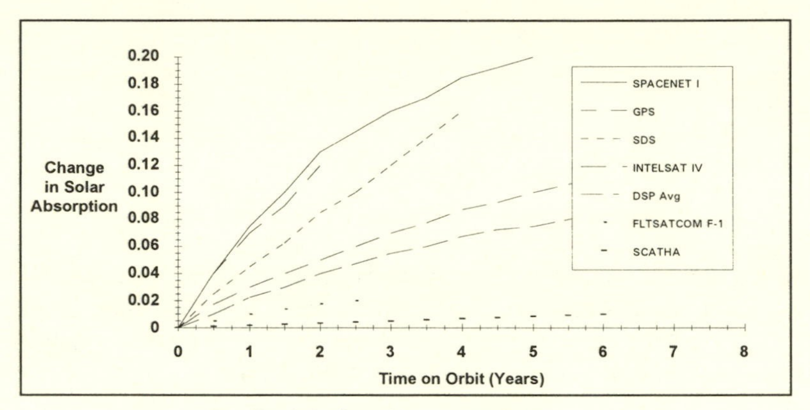

Fig 2.2 On orbit as degradation.
(Courtesy of Dr. Donald F. Gluck, The Aerospace Corporation)

2.2.1 UV Degradation

Only 21% of the Sun's energy passes through the Earth's atmosphere to the surface without being impeded. Thirty-one percent of the Sun's energy is reflected back to space, 29% is scattered down to Earth, and 19% is absorbed as heat. All radiation having a wavelength of less than 0.3 μm, the ultraviolet (UV), is absorbed by the Earth's ozone layer, (fig. 2.3). On orbit, spacecraft surfaces will be exposed to the full strength of the solar UV radiation. The energy of a single photon is related to its wavelength λ or frequency ν according to the relation

$$E = h\nu = \frac{hc}{\lambda}, \tag{2.5}$$

where h is Planck's constant and c is the speed of light. As shown in table 2.2, the energy carried by a single photon of UV light is sufficient to sever many types of organic chemical bonds. As a result, the physical characteristics of many materials change upon exposure to UV radiation. (This is also the reason for the concern over a potential increase in skin cancer associated with the thinning of the Earth's ozone layer by fluorocarbon emissions.) In general, it is possible to group materials into two categories:

"space stable" and "non-space stable." As the name implies, space stable materials show essentially no degradation upon exposure to the orbital environment. The non-space stable materials may be susceptible to a variety of degrading factors. On orbit, it is difficult to separate UV damage from other damage mechanisms, but laboratory tests confirm that many materials may experience a change in α_s on the order of 0.01 or greater during typical spacecraft lifetimes. In many cases, this degradation can be eliminated by choosing only space stable materials for flight.

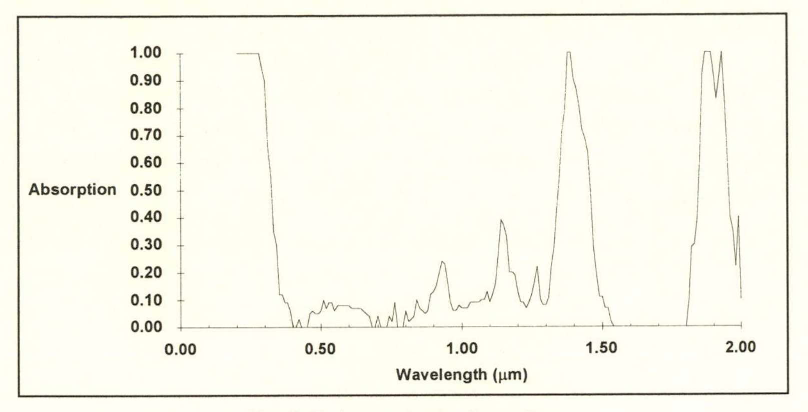

Fig. 2.3 Atmospheric absorption.

Table 2.2
Chemical Bond Energy

Chemical	Bond	Bond Energy at 25° C (kcal/mole)	(eV)	Wavelength (μm)
C-C	Single	80	3.47	0.36
C-N	"	73	3.17	0.39
C-O	"	86	3.73	0.33
C-S	"	65	2.82	0.44
N-N	"	39	1.69	0.73
O-O	"	35	1.52	0.82
Si-Si	"	53	2.30	0.54
S-S	"	58	2.52	0.49
C-C	Double	145	6.29	0.20
C-N	"	147	6.38	0.19
C-O	"	176	7.64	0.16
C-C	Triple	198	8.59	0.14
C-N	"	213	9.24	0.13
C-O	"	179	7.77	0.16

An excellent example of materials degradation upon exposure to the space environment is the change in the properties of the Betacloth liner used in the bay of the space shuttle. Betacloth is a woven fabric of glass fibers that is dipped in an emulsion of TFE Teflon and Polysiloxane during processing. The Teflon is needed to protect the fibers, which would otherwise rub against one another and deteriorate, while the polysiloxane makes the final product more workable. During a one-week shuttle mission it is quite common for the Betacloth to darken visibly. This darkening, which is confirmed by laboratory testing (fig. 2.4), is presumably caused when UV light creates "color centers" in the fabric. These color centers are actually locations where oxygen atoms have been removed from the matrix upon exposure to the UV light. The darkening is in rough proportion to the polysiloxane content. Upon return to the ground and exposure to the atmosphere, the oxygen atoms are replaced, the color centers fade, and the cloth "recovers" its original appearance (fig. 2.5).

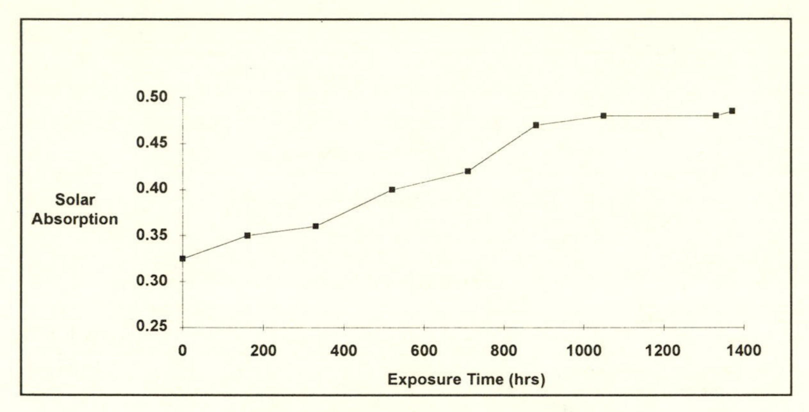

Fig. 2.4 UV degradation of Betacloth.

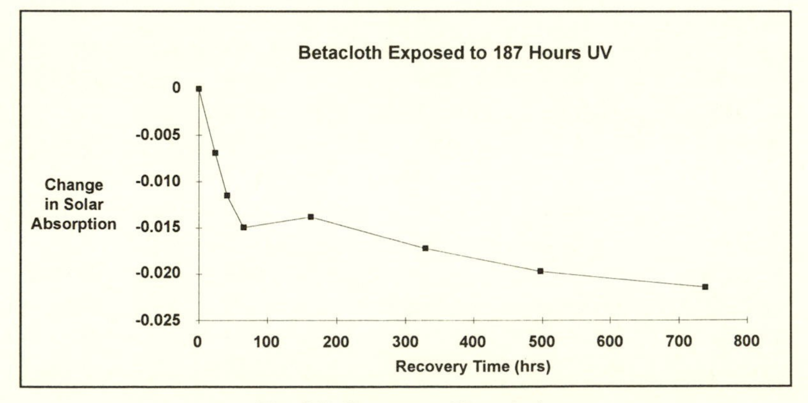

Fig. 2.5 Recovery of Betacloth.

2.2.2 Molecular Contamination

Although some of the degradation noted in figure 2.2 is due to the solar UV, much of the degradation is associated with contamination. Even if the spacecraft surfaces are clean when installed in the launch vehicle, the spacecraft itself may be a source of contamination during launch or on orbit operations through the outgassing process. All but the purest of materials will contain fractional amounts of "volatile" chemicals, either on the surface or dispersed through the material. These volatile chemicals, which may be simply excess chemicals left over from improper catalyst/resin ratios, improper curing, etc., may, over time, migrate to the surface and escape into the local atmosphere. This process is responsible for the familiar odor of plastics or rubber. Once the electrical attraction to the surface of the material has been broken, the outgassed molecules are free to follow ballistic trajectories and may randomly impact other surfaces having a line of sight to the point of departure. Because the deposition occurs one molecule at a time, in thicknesses on the order of Angstroms, this type of contamination is referred to as molecular contamination. If this outgassed material were to build up on a thermal control surface or sensitive optics, the resulting properties could be severely degraded. The surface physics is quite complicated, but some general characteristics of the outgassing process can be noted.

Experimental data indicate that outgassing is seen to vary (1) exponentially as a function of time, (2) inversely as a power of time, or (3) independently of time, depending on the mechanism responsible for the outgassing process. Mechanisms include desorption, diffusion, and decomposition. Desorption is the release of surface molecules that are held by physical or chemical forces. Diffusion is the homogenization that occurs from random thermal motions. Contaminants that diffuse to the surface of a material may have enough thermal energy to escape the surface forces and simply evaporate into the local environment. Finally, decomposition is a type of chemical reaction where a compound divides into two or more simpler substances, which may then outgas through desorption or diffusion. Each of these processes is dependent on a number of factors such as the activation energy, which is a measure of the strength of the surface binding energy, and temperature, which is a measure of the available thermal energy. As shown in table 2.3, each process is characterized by a unique range of activation energies and follows different time dependencies. Because of the lower energy requirement, desorption and diffusion are the primary mechanisms responsible for the outgassing process. In general, desorption is the mechanism responsible for removing contaminant layers from metals. Because desorption will only remove material from the surface, there will generally be little total mass lost due to this process. Diffusion is the mechanism responsible for outgassing from organic materials and, since diffusion involves the total mass of organic material present, the total mass lost due to this process is typically much greater.

Table 2.3
Characteristics of Outgassing Mechanisms

Mechanism	Activation Energy (kcal/mole)	Time Dependence
Desorption	1–10	t^{-1} to t^{-2}
Diffusion	5–15	$t^{-1/2}$
Decomposition	20–80	n/a

The amount of mass loss due to diffusion can be represented by the relation

$$\frac{dm}{dt} = \frac{q_o \exp^{-E_a/RT}}{\sqrt{t}}, \tag{2.6}$$

where q_o is a reaction constant that must be experimentally determined, E_a (kcal/mole) is the activation energy, R (kcal K/mole) is the gas constant, and T (K) is the temperature. Integrating equation 2.6 provides an expression for the amount of mass outgassed between time t_1 and t_2, which is

$$\Delta m = 2q_o \exp^{-E_a/RT}\left(t_2^{1/2} - t_1^{1/2}\right). \tag{2.7}$$

The amount of matter that is outgassed by a material is dependent on the material's specific outgassing characteristics, which are contained in the reaction constant q_o. A standard test of a material's outgassing characteristics, which can be used to determine q_o, is ASTM E 595. In this test, a sample of the material being studied is held at a temperature of 125° C for 24 hours at a pressure of less than 7×10^{-3} Pa. Comparing the initial and final mass of the sample yields the Total Mass Loss (TML). During the outgassing test a collecting plate is held at 25° C in order to measure the amount of Collected Volatile Condensable Material (CVCM). Typical pass/fail criteria for most spacecraft materials are 1% TML and 0.1% CVCM. A third parameter, Water Vapor Regained (WVR), is also measured by subjecting a post-test sample to a 50% relative humidity environment at 23° C for 24 hours. The mass gain is used to infer WVR. Outgassing parameters and activation energies for several typical spacecraft materials are shown in table 2.4.[6-9] The mass density of outgassed contaminants is typically on the order of 1 g/cm^3.

Table 2.4
Outgassing Parameters for Typical Spacecraft Materials

Material	Activation Energy (kcal/mole)	TML (%)	CVCM (%)
Adhesives			
Ablebond 36-02	16.2	0.19	0.00
RTV 566	n/a	0.10	0.02
Scotchweld 2216	11.3	1.25	0.08
Solithane 113/300	12.6	0.66	0.04
Trabond BB-2116	7.96	1.01	0.05
Conformal Coatings			
Epon 815/V 140	31.2	1.07	0.10
Films/Sheet Materials			
Kapton H	n/a	0.77	0.02
Paints/Lacquers/Varnishes			
Cat-a-lac 463-3-8	12.4	2.14	0.03
Chemglaze Z-306	17.2	1.12	0.05
S13GLO	n/a	0.54	0.10
Z-93	n/a	2.54	0.00
Zinc Orthotitinate (ZOT)	n/a	2.48	0.00

If an outgassed molecule impacts a surface, experimental evidence confirms that in most cases the outgassed molecule will not scatter elastically, but will adhere to the surface and establish thermal equilibrium. The contaminant molecule will remain attached to the surface until, following the random probabilities of quantum mechanics, it acquires enough energy to escape the electrical attraction to the surface. The average residence time on the surface is therefore related to the surface temperature and is approximated by

$$\tau(T) = \tau_o \exp^{E_a/RT}, \tag{2.8}$$

where $\tau_o \sim 10^{-13}$ s is a typical oscillation period of the molecule on the surface. As shown in figure 2.6, most outgassed contaminants will have a very short residence time on all but cryogenically cooled surfaces. For example, water, with an activation energy of about 17 kcal/mole, has a residence time of 10^{24} s on a surface at 100 K, but only 0.25 s on a 300 K surface.

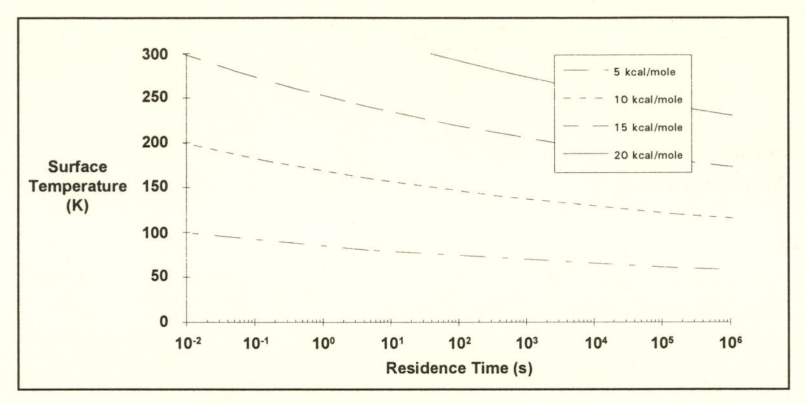

Fig. 2.6 Residence time of contaminants.

A contaminant layer may build up on a surface provided that the arrival rate of contaminants exceeds the rate of departure. That is, contamination will accumulate if at least some of the incident contaminant molecules have a residence time that is long in comparison to the time period of interest. The accumulation rate is approximated by

$$X(t,T) = \gamma(T)\phi(t,T), \qquad (2.9)$$

where γ (T) is the sticking coefficient, i.e., the fraction of incident molecules that attach "permanently" to the surface, and ϕ (μm/s) is the arrival rate. γ may be assumed to be 1.0 for worst-case predictions or for cryogenic surfaces where the residence time of most contaminants is long. However, the ASTM E 595 results would predict a sticking coefficient of 0.1 for room temperature surfaces, in agreement with the fraction of TML that remains as CVCM. If more detailed calculations are required, the evaporation rate can be estimated from the accumulation rate and the residence time.

2.2.2.1 Line of Sight Transport

The arrival rate of contaminants at a given point is dependent on the rate of outgassing from all potential sources and the physical geometry of the point in question relative to each source. In general, the arrival rate is the product of the rate at which mass leaves the source, which can be calculated from equation 2.6, and a geometrical view factor, which is simply a measure of the fraction of matter that leaves the source and impacts a given point of interest. The outgassing view factor bears a strong resemblance to the view factor, or angle factor, used in calculations of radiative heat balance. Consider a plate of area dA_1 which is radiating heat to space with intensity I_1 (W/m^2). What fraction of this heat will impact a surface dA_2 located a distance r from the first plate in the relative orientation shown in figure 2.7?

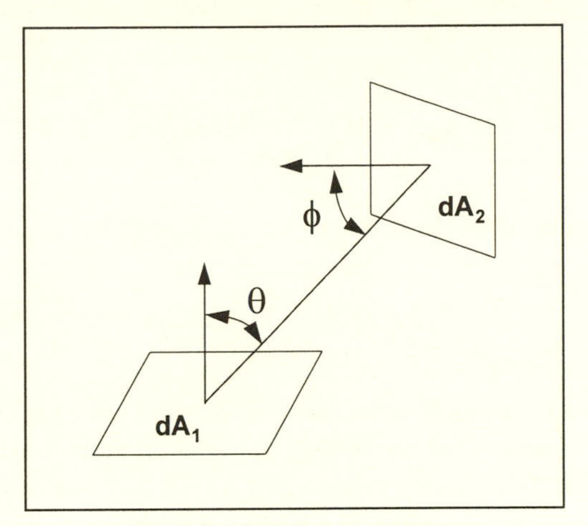

Fig. 2.7 View factor geometry.

The rate at which energy leaves dA_1 in the direction of dA_2 is

$$\Delta q_1 = I_1 \cos\theta dA_1, \tag{2.10}$$

where θ is the angle between the normal to dA_1 and the radius r connecting the center of the plates. The amount of energy that dA_2 intercepts is related to the solid angle that dA_2 subtends, with dA_1 as its origin, which is

$$d\omega_2 = \frac{\cos\phi dA_2}{r^2}. \tag{2.11}$$

Consequently, the amount of energy that leaves dA_1 and impinges dA_2 is

$$\Delta q_{12} = I_1 \frac{\cos\theta\cos\phi}{r^2} dA_1 dA_2. \tag{2.12}$$

If one considers the case of dA_1 radiating over an entire hemisphere, it can be seen that $\cos\phi = 1$ and

$$dA_2 = 2\pi r^2 \sin\theta d\theta. \tag{2.13}$$

Substituting equation 2.13 into 2.12 and performing the integration, it can be shown that the relationship between radiation intensity and emissive power is

$$E = \pi I . \tag{2.14}$$

Consequently, in terms of the power, or strength, of the radiating source, equation 2.12 reduces to

$$\Delta q_{12} = E_1 \frac{\cos\theta\cos\phi}{\pi r^2} dA_1 dA_2 . \tag{2.15}$$

The radiative view factor, from dA_1 to dA_2, is defined by

$$F_{12} = \iint \frac{\cos\theta\cos\phi}{\pi r^2} dA_1 dA_2 . \tag{2.16}$$

This is the view factor used in a variety of thermal modeling codes, such as TRASYS. Integrating over dA_1 and dA_2 provides a view factor between two surfaces. For most outgassing calculations a view factor is needed between a surface, the outgassing source, and a point, the collector. This simplifies the calculation of view factor, which reduces to

$$VF = \int \frac{\cos\theta\cos\phi}{\pi r^2} dA , \tag{2.17}$$

where θ is the angle between the normal to the outgassing source and the radius vector to the collection point, ϕ is the angle between the normal to the collection point and the radius vector from the collection point, and r is the distance between source and collector as illustrated in figure 2.7. The integration in equation 2.17 is performed over the area of the outgassing source. Once the various view factors are known, the arrival rate of mass at a given point can be calculated from

$$\phi = \sum_s VF_s \frac{dm_s}{dt} \frac{1}{\rho_s} , \tag{2.18}$$

where dm_s/dt is rate at which mass leaves the source, found from equation 2.6, ρ_s (g/cm^3) is the contaminant mass density, and the sum is over all possible sources. If a source does not have a direct line of sight to the collection point of interest, its view factor for direct deposition is zero.

An outgassing source may be an extended surface, such as a thermal control panel covered with an outgassing paint, or may be quite localized, such as outgassing through a spacecraft vent or from a single electrical component. Contamination may also come from thermal blankets or multilayer insulation.[10,11] If two or more payloads are carried into orbit on the same launch vehicle, one payload may be degraded by contamination from the other payload.[12] Any material that may outgas is a potential source of contamination.

Example 2.1

Estimate the view factor from a circular vent of radius 10 cm to the point on a solar panel that is nearest the vent. The solar panel is attached to the spacecraft by a 1 m long boom that is located 25 cm below the vent.

By examination:

$$r = (1^2 + 0.25^2)^{1/2} = 1.03 \text{ m}$$

$$\cos\theta = 1 \text{ m}/1.03 \text{ m} = 0.97$$

$$\cos\phi = 0.25 \text{ m}/1.03 \text{ m} = 0.24$$

$$dA = \pi\,(0.1 \text{ m})^2 = 0.0314 \text{ m}^2$$

With this geometry the integrated view factor is approximated by

$$VF \sim (\cos\theta \cos\phi\, dA)/(\pi\, r^2)$$

$$VF \sim (0.97)(0.24)(0.0314 \text{ m}^2)/(3.14 \text{ m}^2) = 0.00219$$

2.2.2.2 Non-Line-of-Sight Transport

It is not always necessary for a contaminant source to have a direct line of sight to a sensitive surface in order for the source to contaminate the surface.[13] A source may outgas matter onto an intermediate surface, which will in turn desorb matter onto the surface of concern. Consequently, reflection, or desorptive transfer, may also need to be considered in a comprehensive contamination analysis. Contaminants may also exit a vehicle and be scattered back through collisions with ambient atmospheric molecules. Obviously, this phenomenon is of greater concern in LEO where the atmospheric density is greatest. In the higher orbits it is rare that this process contributes appreciably to end-of-life levels.

As will be discussed in chapter 4, under certain orbital conditions a spacecraft may develop a negative charge due to the ambient plasma environment. If outgassed contaminant molecules become ionized in the region close to the vehicle (the Debye sheath), the molecule may be reattracted to the vehicle by the resultant electric field. As much as 31% of the contamination deposited on the contamination experiment on the Spacecraft Charging at High Altitudes (SCATHA) spacecraft was associated with periods of spacecraft charging.[14] This phenomenon is of greater concern in the higher orbits where the plasma shielding distance, the Debye length, is much larger. In LEO, where the plasma shielding distance is on the order of 1 cm, contaminant molecules have a high probability of escaping the electrical attraction before being ionized. In GEO, where the shielding distances are on the order of meters or tens of meters, reattraction of contamination is of greater concern. It is important to note that reattraction is only possible during periods of time when 1.) the contaminant flux is sunlit, and 2.) the vehicle is charged negatively.

2.2.3 Effects of Contamination

The presence of a thin contaminant film on the surface of a material will alter its solar absorptance according to the relation

$$\alpha_s^x = \frac{\int \left[1 - R_S(\lambda)\exp\left(-2\alpha_c(\lambda)x\right)\right] S(\lambda)\, d\lambda}{\int S(\lambda)\, d\lambda}, \qquad (2.19)$$

where $\alpha_c(\lambda)$ is the absorptance of the contaminant film (in units of m^{-1}), x (m) is the film thickness, and $R_s(\lambda) = 1 - \alpha_s(\lambda)$ is the solar reflectance of the uncontaminated material. Note that the factor of 2 in the exponential term occurs because a photon must traverse the film, be reflected, and traverse the film a second time to avoid being absorbed. As shown in figure 2.8, the contaminant profile from a mixture of typical spacecraft outgassing products is generally more absorptive in the ultraviolet than the infrared. As a result, molecular contamination is of more concern on remote sensing instruments that operate in the UV than it is for instruments that operate in the IR. Note that the data shown in figure 2.8 are for a room temperature mixture of typical spacecraft contaminants. Individual contaminants or contaminant deposition on cryogenic surfaces may have significantly different absorption characteristics.[15] The contaminant layer will increase the absorptance of the surface material and consequently its equilibrium temperature, as shown in figure 2.9. A general rule of thumb for thermal control surfaces is that the numerical value for the change in absorptance is approximately twice the contaminant thickness in microns.

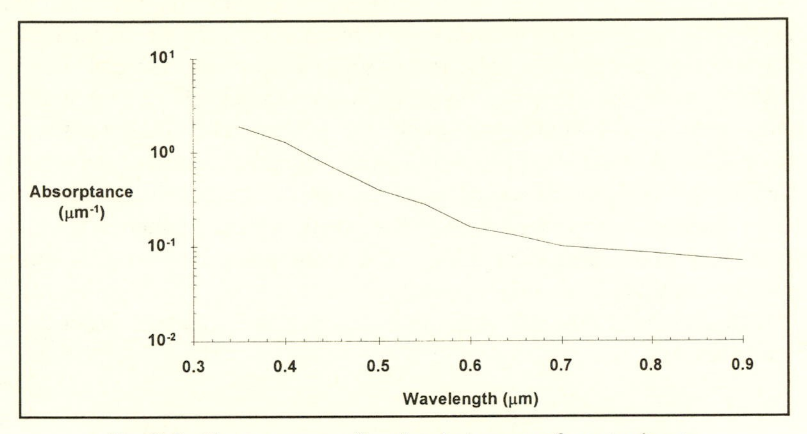

Fig. 2.8 Absorptance profile of typical spacecraft contaminants. (Courtesy of Dr. David F. Hall, The Aerospace Corporation)

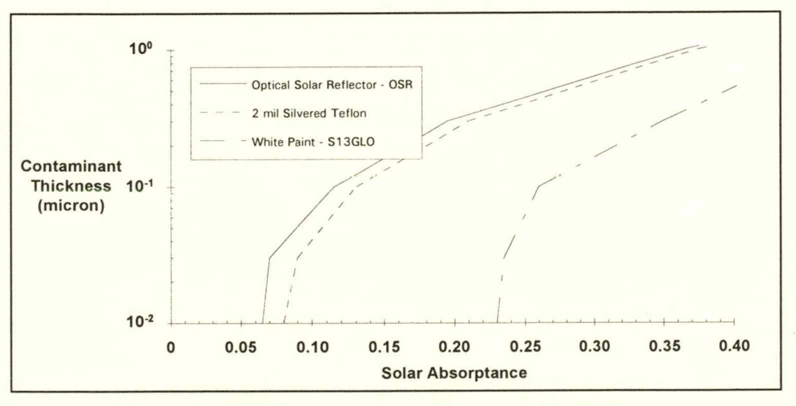

Fig. 2.9 Solar absorptance versus contaminant thickness.

Historically, some spacecraft have had end of life (EOL) α_S values as high as 0.3 – 0.4. For OSRs a typical beginning of life (BOL) value is 0.08. In order to maintain thermal control and still allow for a large degradation in α_S, thermal radiators would have to be significantly oversized at the BOL in order to provide enough radiator surface to do the job at EOL. Oversizing the radiators at BOL would also require the designer to provide significant heater power to offset the increased heat loss at BOL when the α_S value is low. Consequently, minimizing the change in α_S minimizes spacecraft size, weight, and cost. Now that the problem of contamination is more widely understood, EOL α_s values less than 0.2 are achievable with the proper choice of materials and proactive spacecraft design.

In addition to the concern for contamination of thermal control surfaces, there is also the possibility for contamination buildup on optics or solar arrays. The presence of a contaminant film on a lens, mirror, or focal plane will degrade the signal to noise ratio (SNR) of the detector and limit the dynamic range by absorbing light from the target of interest. If the contaminant film becomes too thick the sensor will cease to function properly. For IR sensors, which utilize cryogenically cooled surfaces, the problem may be reversed somewhat if the optics can be warmed up so that the contaminants will evaporate from the surface. Because there is a limit to the number of temperature cycles that a focal plane may survive, this is typically viewed as a last-resort option.

A contaminant film will degrade the output of a solar cell according to the relation

$$F(x) = \frac{\int S(\lambda) I_s(\lambda) \exp(-\alpha_c(\lambda) x) d\lambda}{\int S(\lambda) I_s(\lambda) d\lambda}, \tag{2.20}$$

where $I_s(\lambda)$ (W/m) is the response of the cell, a measure of how effectively the cell converts that color of light into power. A typical solar cell response curve is shown in figure 2.10 and the resulting degradation in cell output due to contamination is shown in figure 2.11. (Note that the final result is slightly dependent on the specific cell design used.) As a rule of thumb, the power output from a cell is decreased by approximately 2% times the contaminant thickness in microns.

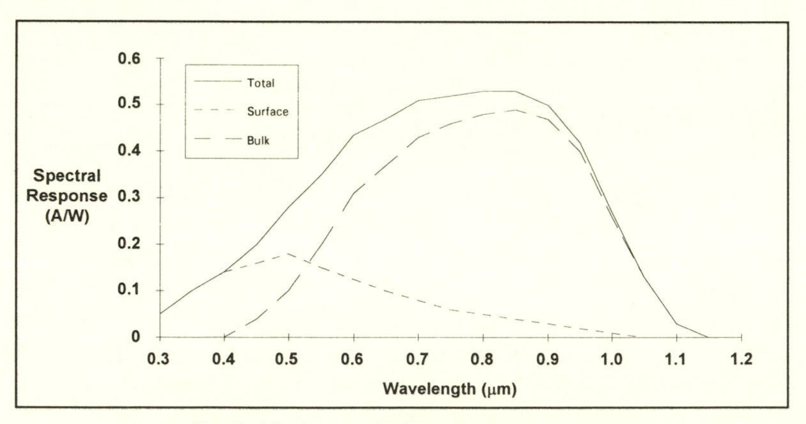

Fig. 2.10 A typical solar cell response curve.

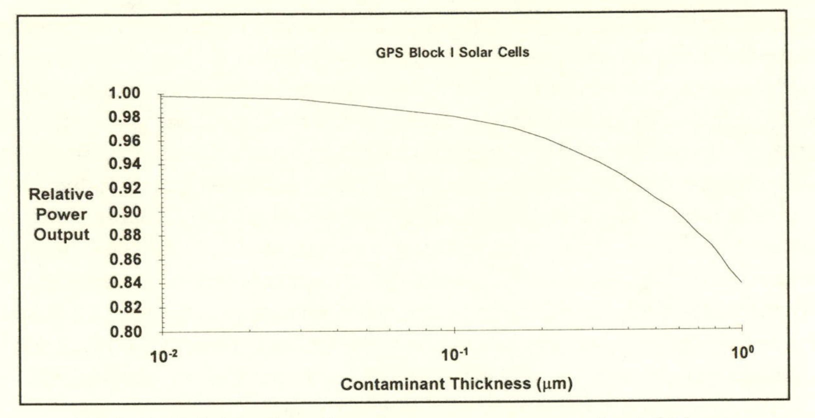

Fig. 2.11 Relative solar cell power output versus contaminant thickness.

Because solar arrays are typically quite warm when illuminated, ~ 60° C, it is appropriate to reexamine the question of sticking coefficients. For room temperature surfaces a sticking coefficient of 0.1 is a typical assumption because it is supported by the ASTM E 595 tests on surfaces at 25° C. The warmer a surface is the less likely contamination is to collect on it due to the decrease in residence time anticipated for impacting molecules. A molecule which would have a residence time of one year on a 25° C surface would have a residence time of about 2.5 days on a surface at 60° C. Consequently, it is appropriate to assume that the sticking coefficient for a solar array will be much less than 0.1, possibly by several orders of magnitude. It is tempting to assume that solar arrays are so warm that no contamination will stick to them. However, as discussed in the next section, there is an abundant amount of experimental evidence that indicates that warm sunlit surfaces are susceptible to contamination.

2.2.4 Synergistic Effects

2.2.4.1 Photochemical Deposition

As with many space environment effects there is often the possibility that synergistic interactions between two or more effects may result in a total degradation that is greater than the sum of its parts. An excellent example is the interaction between the solar UV and molecular contamination. On orbit, illuminated solar arrays would be thought to be too warm to allow for the buildup of molecular contamination due to the very short residence times anticipated for most contaminants. However, it is well documented in the laboratory that the presence of UV light can cause contamination to condense on surfaces that would otherwise remain clean.[16-18] Presumably, the UV light initiates a polymerization process that binds the contaminant molecule to the surface. It is now accepted that this photochemical deposition process was responsible for an accelerated degradation in solar array output noted on the GPS Block I satellites[19] (fig. 2.12). As a result, even warm surfaces may be subject to the deposition of contaminant layers if they are exposed to the solar flux. The rate of photochemically deposited contaminant buildup is seen to increase as the molecular arrival rate decreases (table 2.5). Consequently, even though outgassing rates will decrease with time, the photochemical sticking coefficient and subsequent deposition rates will not decrease at the same rate. As a result, photochemical deposition of contamination may continue to create problems long after outgassing rates have subsided to low values. Power production from the GPS Block I satellites did not become noticeable until after about 3 years on orbit. At this point in a mission the majority of the outgassing has long since ceased and contamination concerns, if not already apparent, have faded.

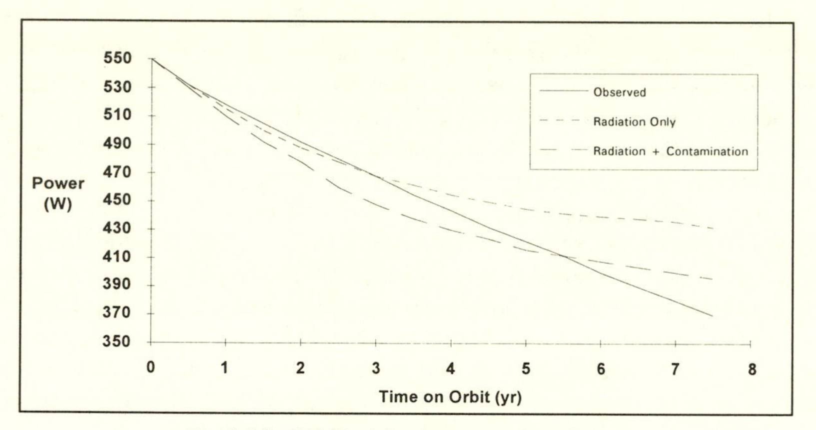

Fig. 2.12 GPS Block I solar power degradation.
(©1991 AIAA—reprinted with permission)

Table 2.5
Photochemical Deposition versus Arrival Rate

Arrival Rate (Ang./hr)	Deposition Rate (Ang./hr)	Sticking Coefficient
5.6	0.112	2.0×10^{-2}
600	0.20	3.3×10^{-4}
14,250	0.55	3.8×10^{-5}

2.2.4.2 Thruster Plume Contamination

Studies of thruster exhaust plumes indicate that thrusters are known to scatter a very small fraction of the ejected mass at angles greater than 90° off of the thruster axis.[20] Typically the amount of mass scattered at the higher angles is less than one part in 10^6, but this is dependent on the specific thruster design. Because such scattering is a real possibility, there is often a concern that firing a spacecraft's propulsion system could scatter contaminants from the exhaust plume onto sensitive surfaces. Hydrazine monopropellant and bipropellant fuels are commonly used for nominal on-orbit station-keeping maneuvers. Both on-orbit measurements and laboratory tests have indicated that hydrazine exhaust does not collect on surfaces warmer than about -45° C.[21,22] Analine impurity decomposition products were witnessed at -101° C, water was collected at -129° C, and ammonia was detected at -167° C. For nominal spacecraft surfaces, deposition from hydrazine thruster plumes will not be a problem. Bipropellant exhaust constitutes a larger contamination concern. The predominant species in the bipropellant plume resulting from incomplete combustion of MMH and N_2O_4 is monomethylhydrazine nitrite (MMH-HNO_3). With an activation energy of 20.48 kcal/mole, MMH-HNO_3 is a contamination concern for cooled surfaces.[23]

2.3 Particulate Contamination

In addition to the molecular contamination just discussed, spacecraft are also subjected to particulate contamination. The term "particulate contamination" refers to the micron-sized pieces of matter that inevitably builds up on exposed surfaces during manufacturing and processing. Individuals who wear eyeglasses are familiar with the fact that they require periodic cleaning to remove dust and other debris that accumulates regularly due to exposure to the environment. Particulates are directly related to air quality on the ground and not the space environment. Consequently, particulate contamination is generally deposited during manufacturing, test, or launch and not during on-orbit operations. However, a brief discussion of the effects of particulate contamination is warranted as many standards and references deal with both molecular and particulate contamination simultaneously. Similarly, the performance of optical elements is related not only to transmission/reflectance criteria, but also to the amount of scattering present in the optics.[24]

The buildup of particulates on a surface is directly related to the amount of particulates in the surrounding air. Viscous drag will balance the fall of particles under the influence of gravity, but over time more and more particulates will fall out of the atmosphere onto exposed surfaces. FED-STD-209E defines air quality in terms of the maximum allowable number of particles per cubic meter, or cubic foot, of air. In SI units, the name of the air class is taken from the base 10 logarithm of the maximum allowable number of particles, 0.5 μm and larger, per cubic meter. In English units, the name of the class is taken from the maximum allowable number of particles, 0.5 μm and larger, per cubic foot. The concentration limits are approximated by

$$\text{particles} / \text{m}^3 = 10^M \left(\frac{0.5\mu\text{m}}{d} \right)^{2.2}$$

$$\text{particles} / \text{ft}^3 = N_c \left(\frac{0.5\mu\text{m}}{d} \right)^{2.2}, \qquad (2.21)$$

where M is the numerical designation of the class in SI units, N_c is the numerical designation of the class in English units, and d (μm) is the particle size. Class limits are illustrated in figure 2.13 and table 2.6. Class M 5.5 (10,000) cleanrooms are typical of most spacecraft manufacturing cleanrooms. Nominal industrial quality air may be class M 8 (3,500,000) or worse, while class M 3.5 (100) laminar flow benches may be required for the assembly of sensitive optical components.

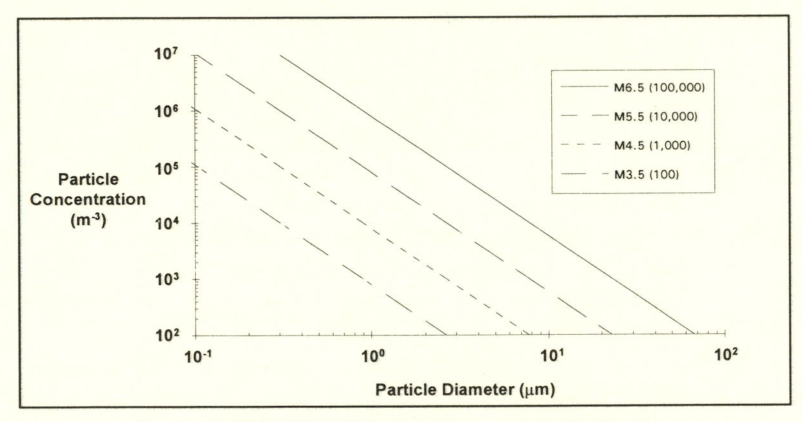

Fig. 2.13 FED STD 209E air volume particle distributions.

Surface cleanliness levels, for both particulates and molecular films, are defined by MIL-STD-1246B. Molecular contaminant films are referred to as Non-Volatile Residue (NVR). NVR is defined as the residual molecular and particulate matter remaining following the filtration of a solvent containing contaminants and evaporation of the solvent at a specified temperature. The term NVR is used as a measure of contamination that can be cleaned from a surface and then fail to evaporate from the solvent mixture used to clean the surface. This is different from the definition for CVCM, which is simply a

Table 2.6.
Air Quality (FED-STD-209E)

Air Class		Class Limits									
		0.1 μm		0.2 μm		0.3 μm		0.4 μm		0.5 μm	
		Volume		Volume		Volume		Volume		Volume	
SI	English	(m^3)	(ft^3)	(m^3)	(ft^3)	(m^3)	(ft^3)	(m^3)	(ft^3)	(m^3)	(ft^3)
M 1		350	9.91	75.7	2.14	30.9	0.875	10.0	0.28	-	-
M 1.5	1	1,240	35.0	265	7.50	106	3.00	35.3	1.00	-	-
M 2		3,500	99.1	757	21.4	309	8.75	100	2.83	-	-
M 2.5	10	12,400	350	2,650	75.0	1,060	30.0	353	10.0	-	-
M 3		35,000	991	7,570	214	3.090	87.5	1,000	28.3	-	-
M 3.5	100	-	-	26,500	750	10,600	300	3,530	100	-	-
M 4		-	-	75,700	2140	30,900	875	10,000	283	-	-
M 4.5	1,000	-	-	-	-	-	-	35,300	1,000	247	7.00
M 5		-	-	-	-	-	-	100,000	2,830	618	17.5
M 5.5	10,000	-	-	-	-	-	-	353,000	10,000	2,470	70.0
M 6		-	-	-	-	-	-	1,000,000	28,300	6,180	175
M 6.5	100,000	-	-	-	-	-	-	3,530,000	100,000	24,700	700
M 7		-	-	-	-	-	-	10,000,000	283,000	61,800	1,750

measure of material collecting on an exposed surface. CVCM is given as a percentage of outgassing mass while NVR is typically specified in terms of mg/ft^2. Similarly, particulate concentrations are measured in terms of particles of size 0.5 μm or greater per ft^2 (table 2.7). (MIL-STD-1246 has yet to be updated to reflect metric units.) The particle concentrations are described by the relation

$$\log(n) = C\left[(\log x_1)^2 - (\log x)^2\right], \tag{2.22}$$

where x (μm) is the particle size, x_1 is the cleanliness level, n is the number of particles/ft^2 greater than or equal to x, and C is a normalization constant numerically approximated by 0.926. The surface cleanliness levels are illustrated in figure 2.14. In reality, the actual distribution of particles on a surface due to particulate fallout from the atmosphere may not agree with MIL-STD-1246B. As shown in table 2.8, the normalization coefficients that are measured in a variety of cleanrooms disagree significantly from the value of 0.926 assumed by the MIL-STD.[25] For uncleaned surfaces a coefficient of 0.383 may agree better with observations. Nevertheless, cleaning does tend to move a surface to a distribution more in agreement with MIL-STD-1246B. Consequently, when using this standard to specify cleanliness levels it is important to limit its use to surfaces that have been cleaned after exposure to fallout.

Table 2.8
Surface Cleanliness Normalization Factors

Source	C
MIL-STD-1246B	*0.926*
NASA/KSC	0.311
Aerospace Corp/KSC	0.380
Martin Marietta/KSC	0.315
TRW Factory	0.354
JPL/Eastern Test Range	0.557
Average slope	*0.383*
Standard deviation	*0.101*

Based on observations of 5 μm-sized particle fallout rates, the average fallout rate of particles onto a horizontal surface, the floor, is given by

$$\frac{dn(t)}{dt} = pN_c^{\,0.773}, \tag{2.23}$$

Table 2.7
MIL-STD-1246B Surface Cleanliness Levels

Particulates			Molecular	
Cleanliness Level	Particulate Size (μm)	Quantity of Particulates (ft^{-2})	Level	NVR (mg/ft^2)
1	1	1	A	< 1.0
5	5	1	B	1.0–2.0
10	5	3	C	2.0–3.0
	10	1		
25	5	23	D	3.0–4.0
	15	3		
	25	1		
50	5	165	E	4.0–5.0
	15	25		
	25	7		
	50	1		
100	15	265	F	5.0–7.0
	25	78		
	50	11		
	100	1		
200	15	4190	G	7.0–10.0
	25	1240		
	50	170		
	100	16		
300	25	7450	H	10.0–15.0
	50	1020		
	100	95		
	250	2		
500	50	11,800	J	15.0–25.0
	100	1100		
	250	26		
	500	1		
750	100	8900		
	250	210		
	500	7		
	750	1		
1000	250	1020		
	500	40		
	750	5		
	1000	1		

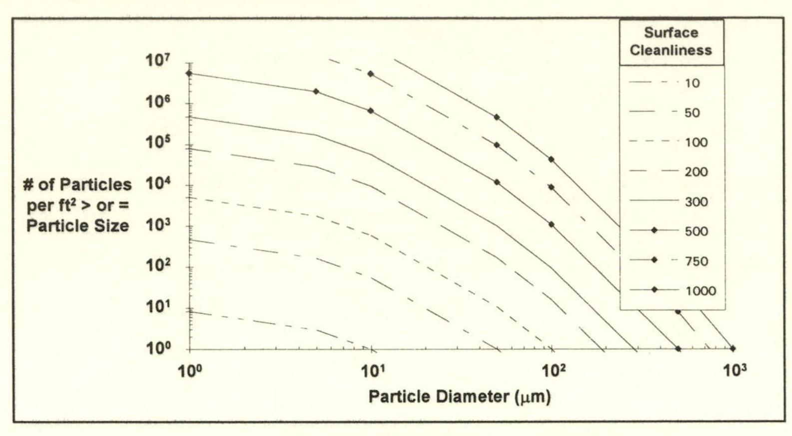

Fig. 2.14 MIL STD 1246B surface particle distributions.

where p is a normalization constant and dn/dt, the fallout rate, is interpreted as the number of particles > 5 μm settled/ft^2/24 hr.[26] Suggested values for p as a function of cleanroom characteristics are listed in table 2.9. Particulate buildup on vertical surfaces should be about 1/10 of the horizontal value while downward facing surfaces may see a buildup of only 1/100 the horizontal value. Integrating equation 2.23 gives the total number of particles present on a surface as a function of time

$$n(t) = pN_c^{0.773}t. \tag{2.24}$$

Inserting equation 2.24 into equation 2.22 allows one to solve for particle surface level as a function of exposure time in a given air-class environment.

Table 2.9
Particle Fallout Normalization Factors

Air Characteristics	p
Still or low velocity air (< 15 air changes/hr)	28,510
Normal cleanroom (15-20 air changes/hr)	2,851
Laminar flow bench (air velocity > 90 ft/min.)	578

The expression for surface cleanliness is

$$\log\left(pN_c{}^{0.773}t\right) = C\left[\left(\log x_1\right)^2 - \left(\log x\right)^2\right]. \tag{2.25}$$

When x is set to 5 μm it is possible to predict surface cleanliness versus time provided that the air class is known. Performing this substitution gives

$$\log^2 x_1 = \left[\log\left(pN_c^{0.773}t\right)^{1/C} + 0.462\right], \tag{2.26}$$

which is illustrated in figure 2.15. It is important to note that equation 2.26 should be used with caution in defining exposure times for surfaces cleaner than level 100. Because equation 2.23 is based on fallout data for 5 μm-sized particles these results may not extrapolate well to surfaces where micron- or submicron-sized particles become significant.

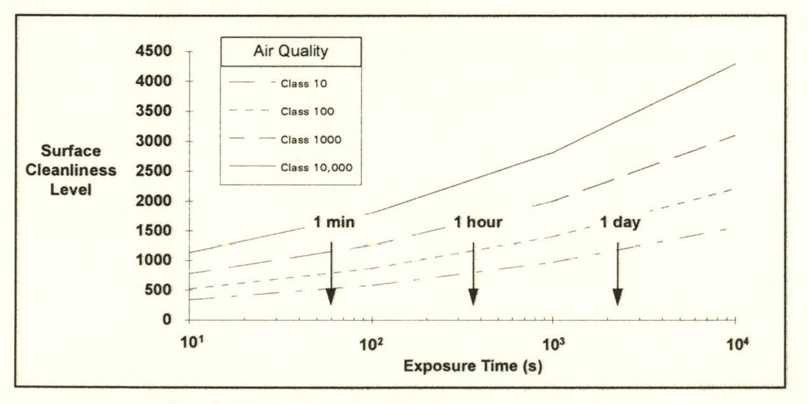

Fig. 2.15 Surface cleanliness versus exposure time.

An example of the magnitude of surface degradation that an optical sensor may encounter during assembly, test, and launch is provided in table 2.10. As shown, good housekeeping practices alone (class 10,000 air) can rarely provide beginning-of-life surface cleanliness values better than level 1000 unless plans are made to clean the surfaces during launch-processing operations. Reducing on-orbit contamination below level 1000 will require stricter attention to detail, such as limiting exposure to class 1000 or better air. Finally, reducing beginning-of-life surface cleanliness below level 500 will require near heroic contamination control measures. As a benchmark, the Hubble Space Telescope requirement limit was level 950.

Table 2.10
Example of Optical Sensor Degradation During Assembly, Testing, Processing and Launch

		Surface Particulate Level		
	Exposure	Air Quality		
Sensor Location	Time	100	1000	10,000
Manufacturing	n/a	100	100	100
Telescope Assembly				
Focal plane integration	1 week	170	363	525
Assembly alignment	2 weeks	220	450	650
Install covers	1 week	235	485	685
Spacecraft Assembly				
Integration	3 months	325	645	900
Test				
Subsystem tests	4 months	380	735	1020
Thermal vacuum tests	1 month	390	750	1040
Final preparations	1 month	395	765	1060
Launch Processing				
Inspection/check out	1 week	398	770	1065
Load propellant	1 week	400	772	1070
Vehicle closeouts	1 week	402	776	1073
Install in launch vehicle	2 weeks	405	782	1081
Ready for launch	1 day	415	784	1082
Launch				
Ascent	10 min	425	790	1082
Initial On-Orbit Checkout				
Instrument deployment	2 weeks	450	800	1100

The degree of obscuration is often used as a measure of surface cleanliness. Because MIL-STD-1246B does not deal explicitly with submicron-sized particles, the resultant predictions of obscuration may be in error if these particulates are not included.[27,28] Visual appearance is a more direct measure of the effect of particle contamination than are the quantitative measurements from MIL-STD-1246B. This is fortunate, in that many NASA procedures for estimating surface cleanliness are based on visual inspection techniques as shown in table 2.11.[29] It is straightforward to verify that an obscuration level great enough to significantly impact a thermal control radiator or a solar array could easily be detected, and eliminated by cleaning, before launch. The concern from particulate contamination is driven primarily by optical instruments. Particulates will scatter incoming light and may create problems for optical devices by limiting the signal to noise ratio or off-axis rejection of the detector. Similarly, particulates may be mistaken by optical devices as false signals. If particulates are ejected from a spacecraft once on orbit, due to solar array deployment, cover ejections, etc., the particulates may remain in the sensor field of view for several seconds and either obscure the intended target or be misinterpreted as star fields, micrometeoroids, etc.

Table 2.11
Visually Clean (VC) Surface Cleanliness Inspection Levels

Level	Illumination (ft candles)	Inspection Distance	Magnification	UV Light	Resolution Limit (μm)
VC 1	50	5–10 ft	1	no	600–1200
VC 1.5	50	2–4 ft	1	no	24–480
VC 2	100	6–18 in	1	no	60–180
VC 3	100–200	6–18 in	2–7	no	10–90
VC 4	100–200	6–18 in	2–7	yes	~ 10

Based on conservation of energy and momentum, a perfectly smooth surface would satisfy the condition that the angle of incidence θ_i is equal to the angle of reflection θ_r. Because no physical surface can ever be perfectly smooth, all real optical devices will have surface imperfections due to cracking, pitting, or particulate contamination. One effect of these imperfections is to scatter a small fraction of the incident light at angles other than $\theta_r = \theta_i$. One measure of the scatter of optical components is the bidirectional reflectance distribution function (BRDF), which is the scattered surface radiance divided by the incident surface radiance.[30] BRDF is a function of many variables and is defined by

$$\mathrm{BRDF} = f_r(\theta_i, \phi_i : \theta_r, \phi_r) = \frac{dL_r(\theta_i, \phi_i : \theta_r, \phi_r : E_i)}{dE_i(\theta_i, \phi_i)}, \tag{2.27}$$

where L_r (W m^{-2} sr^{-1}) is the reflected radiance, E_i (W/m^2) is the incident irradiance, and θ, ϕ are defined as illustrated in figure 2.16. The units of BRDF are sr^{-1}.

BRDF is related to surface cleanliness levels as shown in figure 2.17. BRDF increases as surface contamination levels increase because each particulate is able to scatter light away from the desired angle of reflection.

The four curves illustrated assume light of 10.6 μm wavelength and are based on Mie scattering theory.[31] Note that in order to compare the results of figure 2.17 to other wavebands, it will be necessary to multiply the BRDF values by the factor $(10.6\ \mu m/\lambda_{other})^{2.2}$. Also, more exact calculations of BRDF predict roughly double the value obtained from Mie theory.[32]

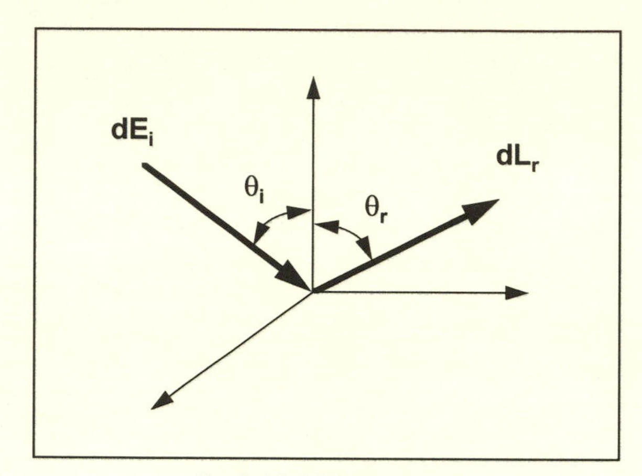

Fig. 2.16 BRDF geometry.

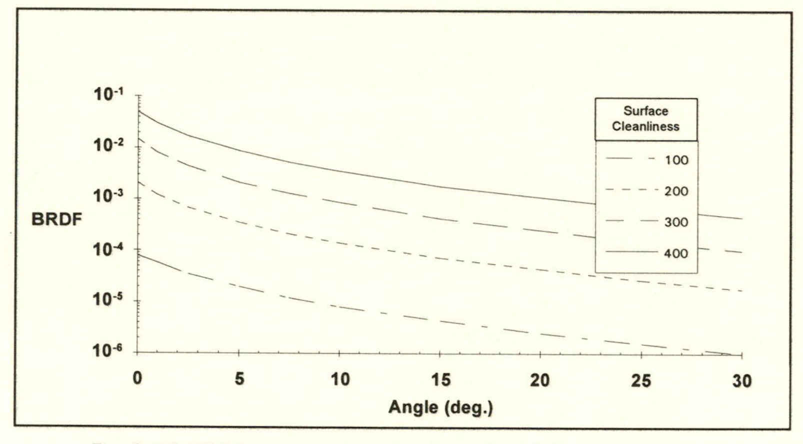

Fig. 2.17 BRDF versus surface cleanliness for 10.7 μm wavelength.

BRDF is closely tied to sensor performance characteristics. For example, one measure of sensor performance is the Point Source Transmittance (PST). PST is defined as the fraction of the signal strength from an off-axis point source that is transmitted to the focus of the optical train. The relation between BRDF and PST is

$$\mathrm{PST} = \frac{\pi}{4\left(\frac{L}{d}\right)^2} \frac{\cos\theta}{\theta^s}\left[1 - \frac{\theta}{\operatorname{atan}\left(\frac{L}{d}\right)}\right]\mathrm{BRDF}, \tag{2.28}$$

where L is the focal length of the optical train, d is the aperture diameter, θ is the angle between the normal and the point source, and s is a parameter that varies from 1, for typical optics, to 2, for superpolished mirrors. Consider the example of a space-borne sensor, such as the Hubble Space Telescope, that is pointing at a faint star cluster that lies within a few degrees of a bright object such as the Sun. The fraction of energy from the Sun that reaches the focal plane will be the product of the total solar output, 1371 W/m^2, and the sensor PST. The PST value, and consequently the BRDF value, would have to be quite small in order for the reflected solar radiation not to overwhelm the faint signal from the star cluster, or possibly even damage the sensor itself. This places a dual constraint on both the Sun exclusion angle (the minimum angular separation between the Sun and objects of interest) and surface cleanliness levels. Because of difficulties associated with manufacturing ideal surfaces, BRDF values for "perfect" surfaces can rarely be less than 10^{-6} at 1°. Machined surfaces may be in the range 10^{-5} to 10^{-3} at 1°, while a complete optical train may be more on the order of 10^{-2} at 1°. Note that the BRDF values required of a completed optical train must be decoupled in order to identify the required surface cleanliness requirement for individual surfaces.

Particulates are deposited on surfaces during ground operations. However, these particulates may be released on orbit by nominal spacecraft operations. On unmanned spacecraft this may occur due to articulation of solar arrays, thermal expansion/contraction, the release of covers, etc. On manned missions, like the shuttle, venting and water dumps may generate particulates. Regardless of the source, particulates released on orbit may interfere with optical measurements. Optical measurements taken by a photometer having a 32° field of view during STS missions 2, 3, 4, and 9 reported particulates during every available viewing opportunity during the first 13 hours of a mission.[33] After about 24 hours on orbit the particulate viewing rate decayed to a quiescent rate of about 500 particles of size > 10 μm per orbit. Other experimenters have reported detecting 1100 particulates from 4 hours of data taken early in the Spacelab 2 mission (STS-54F).[34] The particulates were slow-moving and had temperatures in the range 190 to 350 K. As expected, the size distribution was in agreement with that observed at the shuttle preparation facilities. In support of the shuttle program, studies of micrometeoroid (MM) impact have found that small MMs, which are numerous, are able to dislodge large particulates from surfaces quite easily but do not, in general, remove submicron-sized particles.[35] Larger MMs, which are less frequent, are able to remove both large and small particulates. However, because of the nature of the hypervelocity impact (chapter 6), particulates will be generated by the backsplash of material from the crater produced by the impact. It is predicted that 5.7×10^3 particulates of size ≥ 5 μm would be liberated from the surface of the shuttle each day by MM impacts. Conversely, between 6.9×10^5 and 1.4×10^7 particulates of size ≥ 2 μm, or $2.0 - 3.3 \times 10^5$ particulates ≥ 10 μm in size, would be generated from the crater backsplash. If orbital debris (OD) impacts are factored in, these numbers may increase significantly.

Finally, particulates may also have an impact on surface emittance. The degradation in thermal emittance of the front surface of a contaminated cryogenic mirror is approximately one-half of the surface obscuration of the mirror.[36] This is in essence a statement that dust appears to have 50% of the effect of perfectly "black" dust in perfect contact with the surface. Again, this is primarily of concern for optical surfaces as dust levels sufficient to induce a change in the emittance of a radiator, for example, should be detectable before launch.

2.4 Additional Concerns

2.4.1 Measuring Molecular Film Thicknesses

The procedure used to determine the amount of material remaining on a surface, the non-volatile residue (NVR), is ASTM E 1235. The surface is wiped using Soxhlet-extracted wipers and a solvent mixture of 75% 1,1,1-tricholorethane (methyl chloroform) and 25% ethyl alcohol (ethanol). The NVR is extracted from the wipers with additional solvent, which is either evaporated in a vacuum oven or in a class 100 unidirectional air-flow hood. The mass of the residue minus the mass of a blank sample, divided by the area wiped, is equal to the mass per unit area of NVR on the surface.

2.4.2 Measuring Particulate Levels

The procedure used to determine the distribution of particulates on a surface is ASTM E 1216. Essentially, a tape sample is applied to a surface in order to cause any particulates present to bond to the tape. The tape sample is then removed and examined under a microscope. Provided the sample is large enough to be statistically significant, the results will yield surface cleanliness.

2.4.3 Cleaning Techniques

During ground processing, most molecular films may be removed by chemical wiping methods. Although certain films, such as those containing silicone, for example, may be extremely difficult to remove. Surface inspection is typically used to verify surface cleanliness before launch. On orbit, warming a surface is typically the only option present to remove molecular films.

Chemical wiping will also remove the majority of the particulates resident on a surface. Where optical surfaces are involved, making physical contact with the surface is generally prohibited as this will damage the surface smoothness or optical coatings. A variety of noncontact techniques have been investigated as potential means to clean particulate contamination from surfaces. The forces adhering particulates to a surface are ultimately electrical in nature. In an air environment, the attractive forces between a 1 μm glass particle and a wafer surface are estimated at 71% capillary (0.045 dynes), 22% van der Waals – London (0.014 dynes), 7% electrical double layer (0.003 dynes), and 1% electrostatic image (0.001 dynes).[37] In general, particle adhesive forces vary widely with particle size, shape, and material characteristics. Some particles may fall off under the influence of gravity, while others will remain attached under the influence of 1000 g's. In order to clean particulates from a surface, an external force must be applied to the particulates in order to overcome the adhesive forces. One accepted method is to simply blow air across the surface. Ninety percent cleaning efficiencies associated with the removal of 10 μm-sized particles have been reported for > 150 psi cold gas jets. Larger particles can be removed in this way, but most results indicate that particulates smaller than 0.5 μm are extremely difficult to remove. Ultrasonic and megasonic agitation methods are often used in the semiconductor industry, but these are obviously unsuitable for bulk cleaning of assembled optics. Other methods, which utilize electromagnetic waves, plasmas, and/or electron/ion beams have also been studied.[38] Because of the possibility for recontamination during launch processing, it is common for spacecraft to be transported only after being cleaned, bagged, and connected to a filtered dry nitrogen purging system. Sensitive surfaces may also be stored facing downward to further minimize particulate buildup.

2.5 Space and Ground-Based Testing

2.5.1 Calorimeters

In order to measure degradation of thermal control materials on orbit, spacecraft may be instrumented with devices called *calorimeters*.

Essentially, a calorimeter is a thermistor that is calibrated to operate over the predicted range of temperatures. By isolating a sample material from the spacecraft and allowing it to establish thermal equilibrium, its temperature is indicative of its α_s/ε ratio. Changes in α_s/ε will be indicated by a change in temperature of the sample. If the thermistor has been properly calibrated, the change in α_s can be inferred. The relative uncertainty in absorptance is dependent on the uncertainty in emittance, temperature, solar irradiance, and heat loss due to coupling to the surrounding material.[39] Because of this coupling, the absorptance is given by

$$\alpha_s = \frac{\varepsilon\sigma T^4 + Q_L''}{S}, \tag{2.29}$$

where Q_L'' is the heat loss due to coupling between the sample materials and its surrounding supports. Differentiating equation 2.29 will provide the relative uncertainty in α_s. If preflight calibration is performed, a robust design may be able to infer changes in absorptance as low as 0.0005.

2.5.2 Quartz Crystal Microbalance (QCM)

A calorimeter can measure the change in α_s of a surface, but it cannot determine whether the change is due to the physical deposition of contaminating material on the surface or due to degradation of the sample itself. A Quartz Crystal Microbalance (QCM) may be used to measure the physical mass of a molecular film.[40] Essentially, the QCM operates by comparing the resonant frequencies of two quartz crystals. One crystal is exposed to the environment and the other is shielded. The resonant frequency of the exposed crystal will change if mass is deposited on its surface. Consequently, by examining the change in resonant frequency, mass deposition can be inferred.

2.5.3 Material Outgassing

The standard outgassing test, ASTM E 595, does not yield a unique solution to the question of a material's outgassing properties. The actual time dependence of the outgassing process is undetermined, as are the adhesion characteristics of the outgassed volatiles. More refined tests are required in order to address these questions. One such test has been proposed and is (as of summer 1994) in the process of being reviewed by the ASTM.[41] Rather than use a single collector at a single temperature, the refined test would make use of different detectors at different temperatures. Additional testing can be devised to study the effect of contaminant film darkening in the space environment.[42]

2.6 Design Guidelines

As listed in table 2.12, the spacecraft designer has a variety of tools available to use in developing a spacecraft that is relatively insensitive to UV degradation or contamination. Materials selection is the most obvious first step. Choosing materials that are UV resistant and, at the same time, low outgassing is essential to mission success. Second, the design must accommodate the possibility of self-contamination and should take steps to minimize this occurrence. For example, spacecraft vents or thrusters can be directed away from sensitive surfaces to minimize line-of-sight transport. Thermal control subsystems should contain an adequate margin for degradation during the mission lifetime and may need to allow for the possibility of heating up cryogenic surfaces periodically to allow for boil-off of deposited contaminants. If these steps are not sufficient, many materials may need to be pretreated by thermal baking the materials on the ground to speed up the outgassing rate and minimize the amount that occurs on orbit. Similarly, leaving sensitive optics warm and covered for the first few days on orbit will allow for outgassing to progress without the possibility of contamination.

Table 2.12
Vacuum Environment Effects Design Guidelines

Materials selection	Choose space stable materials and coatings
Configuration	Vent outgassed material away from sensitive surfaces
Margin	Allow for degradation in thermal/optical properties on orbit
Materials pre-treatment	Consider vacuum bakeout of materials before installation in vehicle
Flight & ground operations	Provide time for on orbit bakeout during early operations; provide cryogenic surfaces the opportunity to warm up and outgas contaminant films

2.7 Exercises

1. In order to estimate the number of photons that the Earth receives from the Sun, per unit area, that have sufficient energy to sever a single C-C bond,

   a. Identify the maximum wavelength a photon may have and still be capable of severing the bond.

   b. Make a straightline approximation to $S(\lambda)$ over the waveband of interest.

   c. Integrate the ratio [$S(\lambda)$/*photon energy*] over the waveband of interest.

2. Calculate the view factor from a thermal radiator to points A and B on a solar array as illustrated below by each of three methods:

   a. Calculate the value $\cos\theta\cos\phi/\pi r^2$ between the center of the radiator and points A and B and multiply this value by the area of the radiator.

   b. Treat the radiator as 8 equal mini-radiators and calculate the view factor for each mini-radiator as in (a).

   c. Obtain a closed-form algebraic solution for the view factor. (Hint: Define one axis of the coordinate system to be parallel to the solar panel boom.)

   d. Discuss the pros and cons of each method.

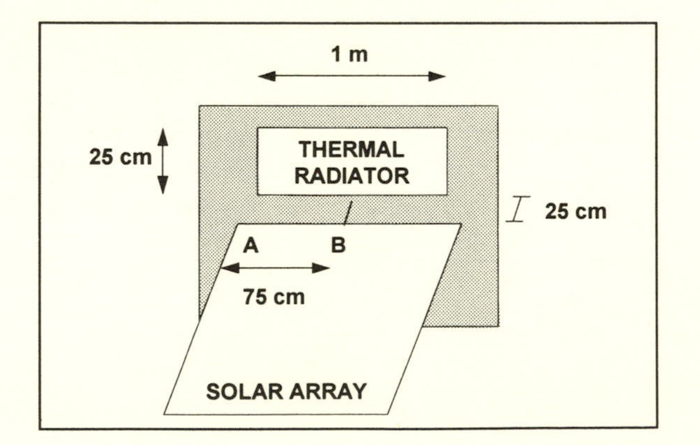

3. Derive a closed-form expression for the view factor from a cylindrical telescope baffle of length L and diameter D to the center of an optical element (the mirror), as shown below.

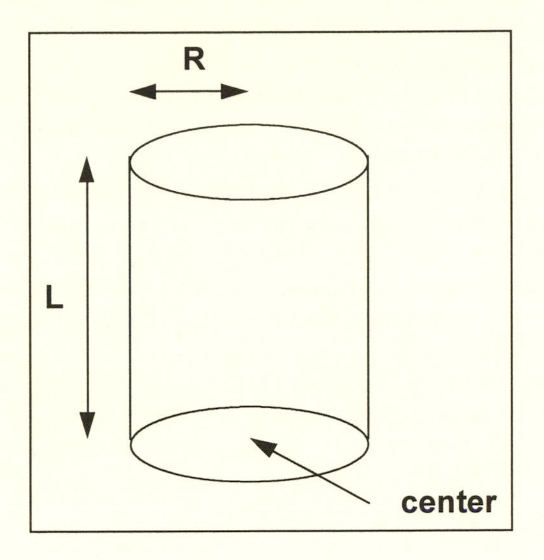

4. Calculate the amount of mass that will be outgassed each day during the first week on orbit by a 10 kg source with E_a = 10 kcal/mole that meets the ASTM E 595 requirements and is held at a temperature of (a) O° C, (b) 25° C, (c) 50° C.

5. If the mass outgassed in problem 4(b) is directed to exit through a circular vent of radius 2 cm having a view factor of 1×10^{-4} to the closest point of a shaded thermal control surface at 25° C, what is the deposition of material on this point as a function of time? Assume the density of contaminants is 1 g/cm^3.

6. If the thermal control surface in problem 5 is replaced with an illuminated solar panel at 60° C, what is the deposition of material as a function of time assuming the following:

   a. A constant sticking coefficient (state and justify your assumption).

   b. A sticking coefficient that varies with arrival rate.

2.8 Applicable Standards

ASTM E 595, *Total Mass Loss and Collected Volatile Condensable Materials from Outgassing in a Vacuum Environment*, 28 November 1990.

ASTM E 1216, *Standard Practice for Sampling for Surface Particle Contamination by Tape Lift*, 27 November 1987.

ASTM E 1234, *Standard Test Method for Gravimetric Determination of Nonvolatile Residue (NVR) in Environmentally Controlled Areas for Spacecraft*, 25 March 1988.

ASTM F 24, *Standard Method for Measuring and Counting Particulate Contamination on Surfaces*, 31 August 1965. (Reapproved 1983).

ASTM F 25, *Standard Test Method for Sizing and Counting Airborne Particulate Contamination in Clean Rooms and Other Dust-Controlled Areas Designed for Electronic and Similar Applications*, 15 October 1968. (Reapproved 1988).

FED STD 209E, *Airborne Particulate Cleanliness Classes in Cleanrooms and Clean Zones*, 11 September 1992.

MIL HDBK 406, *Contamination Control Technology Cleaning Materials for Precision Precleaning and use in Clean Rooms and Clean Work Stations*, 14 June 1988.

MIL STD 1246B, *Product Cleanliness Levels and Contamination Control Program*, 4 September 1987.

2.9 References

1. Hall, D. F., and Fote, A. A., "α_s/ε_H Measurements of Thermal Control Coatings on the P78-2 (SCATHA) Spacecraft," in *Heat Transfer and Thermal Control*, ed. A. L. Crosbie, Vol. 78, p. 467, Progress in Aeronautics and Astronautics (1981).
2. Pence, W. R., and Grant, T. J., "α_S Measurements of Thermal Control Coatings on Navstar Global Positioning System Spacecraft," in *Spacecraft Radiative Transfer and Temperature Control*, ed. T. E. Horton, Vol. 83, p. 234, Progress in Astronautics and Aeronautics (1984).
3. Naumann, R. J., "Skylab-Induced Environment," in *Scientific Investigations of the Skylab Satellite*, ed. M. I. Kent, E. Stuhlinger, and S. T. Wu, Vol. 48, p. 383, Progress in Astronautics and Aeronautics (1974).
4. Mossman, D. L., Bostic, H. D., and Carlos, J. R., "Contamination Induced Degradation of Optical Solar Reflectors in Geosynchronous Orbit," in Society of Photo Optical Instrumentation Engineers, *Optical*

System Contamination Effects, Measurement, Control, Vol. 777, p. 12 (1987).

5. Ahern, J. E., Belcher, R. L., and Ruff, R. D., "Analysis of Contamination Degradation of Thermal Control Surfaces on Operational Satellites," paper 83-1449, American Institute of Aeronautics and Astronautics, 18th Thermophysics Conference, Montreal, Canada (1983).
6. Campbell, W. A., Jr., and Scialdone, J. J., *Outgassing Data for Selecting Spacecraft Materials*, NASA Reference Publication 1124, Rev. 3 (1993).
7. Scialdone, J. J., "An Estimate of the Outgassing of Space Payloads and Its Gaseous Influence on the Environment," *J. Spacecraft*, 23, no. 4, p. 373 (1986).
8. Scialdone, J. J., "Characterization of the Outgassing of Spacecraft Materials," in Society of Photo Optical Instrumentation Engineers, *Shuttle Optical Environment*, Vol. 287, p. 2 (1981).
9. Muscari, J. A., and O'Donnell, T., "Mass Loss Parameters for Typical Shuttle Materials," in Society of Photo Optical Instrumentation Engineers, *Shuttle Optical Environment*, Vol. 287, p. 20 (1981).
10. Glassford, A. P. M., "Outgassing Behavior of Multilayer Insulation Materials," *J. Spacecraft*, 7, no. 12, p. 1464 (1970).
11. Scialdone, J. J., "Spacecraft Thermal Blanket Cleaning: Vacuum Baking or Gaseous Flow Purging," *J. Spacecraft*, 30, no. 2, p. 208 (1993).
12. Jeffrey, J. A., Maag, C. R., Seastrom, J. W., and Weber, M. F., "Assessment of Contamination in the Shuttle Bay," in Society of Photo Optical Instrumentation Engineers, *Shuttle Optical Environment*, Vol. 287, p. 41 (1981).
13. Alan Kan, H. K., "Desorptive Transfer: A Mechanism of Contamination Transfer in Spacecraft," *J. Spacecraft*, 12, no. 1, p. 62 (1975).
14. Hall, D. F., and Wakimoto, J. N., "Further Flight Evidence of Spacecraft Surface Contamination Rate Enhancement by Spacecraft Charging," paper 84-1703, American Institute of Aeronautics and Astronautics, 19th Thermophysics Conference, Snowmass, CO (1984).
15. Wood, B. E., Bertrand, W. T., Bryson, R. J., Seiber, B. L., Falco, P. M., and Cull, R. A., "Surface Effects of Satellite Material Outgassing Products," *J. Thermophysics*, 2, no. 4, p. 289 (1988).
16. Hall, D. F., Stewart, T. B., and Hayes, R. R., "Photo-Enhanced Spacecraft Contamination Deposition," paper 85-0953, American Institute of Aeronautics and Astronautics, 20th Thermophysics Conference, Williamsburg, VA (1985).
17. Stewart, T. B., Arnold, G. S., Hall, D. F., Marvin, D. C., Hwang, W. C., Young Owl, R. C., and Marten, H. D., "Photochemical Spacecraft Self-Contamination: Laboratory Results and Systems Impacts," *J. Spacecraft*, 26, no. 5, p. 358 (1989).

18. Stewart, T. B., Arnold, G. S., Hall, D. F., and Marten, H. D., "Absolute Rates of Vacuum Ultraviolet Photochemical Deposition of Organic Films," *J. Phys. Chem.*, Vol. 93, p. 2393 (March 1989).
19. Tribble, A. C., and Haffner, J. W., "Estimates of Photochemically Deposited Contamination on the GPS Satellites," *J. Spacecraft*, 28, no. 2, p. 222 (1991).
20. Simmons, G. A., "Effect of Nozzle Boundary Layers on Rocket Exhaust Plumes," *AIAA Journal*, 10, no. 11, p. 1534 (1972).
21. Fote, A. A., and Hall, D. F., "Contamination Measurements during the Firing of the Solid Propellant Apogee Insertion Motor on the P78-2 (SCATHA) Spacecraft," in Society of Photo Optical Instrumentation Engineers, *Shuttle Optical Environment*, Vol. 287, p. 95 (1981).
22. Carre, D. J., and Hall, D. F., "Contamination Measurements during Operation of Hydrazine Thrusters on the P78-2 (SCATHA) Satellite," *J. Spacecraft*, 20, no. 5, p. 444 (1983).
23. Liu, Chang-Keng, and Glassford, A. P. M., "Contamination Effect of MMH/N_2O_4 Rocket Plume Product Deposition," *J. Spacecraft*, 18, no. 4, p. 306 (1981).
24. Bremer, J. C., "General Contamination Criteria for Optical Surfaces," in Society of Photo Optical Instrumentation Engineers, *Shuttle Optical Environment*, Vol. 287, p. 10 (1981).
25. Hamberg, O.,"Particle Fallout Predictions for Clean Rooms," *J. Env. Sci.*, 25, no. 3, p. 15 (1982).
26. Buch, J. D., and Barsh, M. K., "Analysis of Particulate Contamination Buildup on Surfaces," in Society of Photo Optical Instrumentation Engineers, *Optical System Contamination: Effects, Measurement, Control*, Vol. 777, p. 43 (1987).
27. Kelley, J. G., "Measurement of Particulate Contamination," *J. Spacecraft*, 23, no. 6, p. 641 (1986).
28. Barengoltz, J. B., "Calculating Obscuration Ratios of Contaminated Surfaces," *NASA Tech Briefs*, 13, no. 8, item 2 (1989).
29. *Specifications–Contamination Control Requirements for the Space Shuttle Program*, NASA-SN-C-0005, Rev. A (January 1982).
30. Nicodemus, F. E., Richmond, J. C., Hsia, J. J., Ginsberg, I. W., and Limperis, T., *Geometrical Considerations and Nomenclature for Reflectance*, National Bureau of Standards Monograph 160, PB 273 439 (October 1977).
31. Young, R. P., "Low-Scatter Mirror Degradation by Particle Contamination," *Opt. Eng.*, 15, no. 6, p. 516 (1976).
32. Johnson, B. R., "Exact Calculation of Light Scattering from a Particle on a Mirror," in Society of Photo Optical Instrumentation Engineers, *Optical System Contamination*, Vol. 1754, p. 72 (1992).

33. Clifton, K. S., and Owens, J. K., "Optical Contamination Measurements on Early Shuttle Missions," *App. Optics*, 27, no. 3, p. 603 (1988).
34. Simpson, J. P., Witteborn, F. C., Graps, A., Fazio, G. G., and Koch, D. G., "Particle Sightings by the Infrared Telescope on Spacelab 2," *J. Spacecraft*, 30, no. 2, p. 216 (1993).
35. Barengoltz, J., "Particle Release Rates from Shuttle Orbiter Surfaces Due to Meteoroid Impact," *J. Spacecraft*, 17, no. 1, p. 58 (1980).
36. Facey, T. A., and Nonnenmacher, A. L., "Measurement of Total Hemispherical Emissivity of Contaminated Mirror Surfaces," in Society of Photo Optical Instrumentation Engineers, *Stray Light and Contamination in Optical Systems*, Vol. 967, p. 308 (1988).
37. Khilnani, A., and Matsuhiro, D., "Adhesion Forces in Particle Removal From Wafer Surfaces," *Microcontamination*, p. 28 (April 1986).
38. Feicht, J. R., Blanco, J. R., and Champetier, R. J., "Dust Removal From Mirrors: Experiments and Analysis of Adhesion Forces," in Society of Photo Optical Instrumentation Engineers, *Stray Light and Contamination in Optical Systems*, Vol. 967, p. 19 (1988).
39. Brosmer, M. A., Fischer, W. D., and Hall, D. F., *Thermal Analysis of Flight Calorimeter Instrument Designs and Calibration Test Methods*, paper 87-1622. American Institute of Aeronautics and Astronautics, 22d Thermosphyics Conference, Honolulu, HI (8-10 June 1987).
40. Wallace, D. A., and Wallace, S. A., *Realistic Performance Specifications for Flight Quartz Crystal Microbalance Instruments for Contamination Measurement on Spacecraft*, paper 88-2727, American Institute of Aeronautics and Astronautics, Thermophysics, Plasmadynamics, and Lasers Conference, San Antonio, TX (June 1988).
41. Glassford, A. P. M., and Garrett, J. W., *Characterization of Contamination Generation Characteristics of Satellite Materials*, WRDC-TR-89-4114, Wright-Patterson Air Force Base, Dayton, OH, Air Force Systems Command, 22 November 1989.
42. Judeikis, H. S., Arnold, G. S., Hill, M., Young Owl, R. C., and Hall, D. F., "Design of a Laboratory Study of Contaminant Film Darkening in Space," in Society of Photo Optical Instrumentation Engineers, *Scatter from Optical Components*, Vol. 1165, p. 406, (1989).

3 The Neutral Environment

The empty, vast, and wandering air.
—Shakespeare, Richard III

3.1 Overview

Although it is far too tenuous to support human life, in LEO the neutral atmosphere is of sufficient density to cause significant interactions with a spacecraft moving at orbital velocities of 8 km/s. The impact of atoms at these high speeds will give rise to an aerodynamic drag force and may physically sputter material from some surfaces. We will see shortly that the most abundant element in LEO is atomic oxygen which, due to its highly reactive nature, may chemically erode surfaces or give rise to a visible glow that may interfere with remote sensing observations. The study of the neutral atmosphere is often referred to as *aeronomy*.[1-3] The science of aeronomy has developed its own special nomenclature for the various regions, or layers, of the Earth's atmosphere depending on the type of physical processes responsible for the phenomena observed there. Each of the regions, or altitude bands, are referred to by the term "sphere." The upper boundary to the spheres is given the term "pause," although typically the boundaries are not well defined and may vary over many tens of kilometers in altitude. Figure 3.1 illustrates the various regions as a function of altitude.

The layer of the atmosphere closest to the surface of the Earth is called the *troposphere*. It is characterized by a roughly uniform composition: 78% N_2, 21% O_2, 1% argon, and a scattering of other elements (table 3.1). In the troposphere the density and temperature decrease as altitude increases. Above the troposphere lies the stratosphere. The *stratosphere* begins at about 11–12 km altitude and is given a separate terminology because of early observations that indicated a nearly isothermal region of temperatures at higher altitudes. This temperature increase is due to the absorption of UV rays by ozone in the 10–80 km altitude band (fig. 2.3). The upper boundary of the stratosphere is not well defined, but at about 45 km the stratopause gives way to a region of colder temperatures called the *mesosphere.* The mesosphere continues upward to about 80–85 km, where the atmosphere is at its coldest, about 180 K. Because the mesosphere lies above the highest altitudes reachable by balloons and below the lowest perigee of spacecraft, it is the least studied region of the atmosphere and is sometimes jokingly referred to as the "ignoro-sphere."

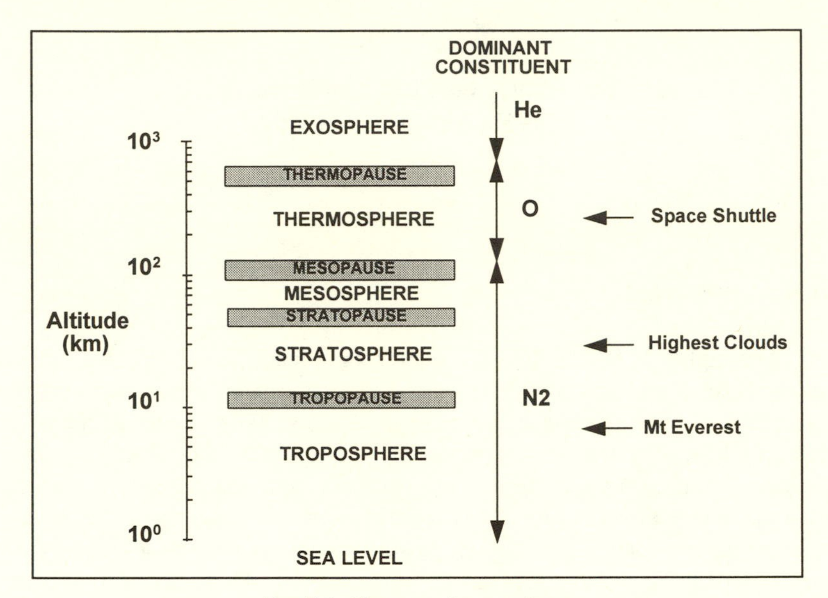

Fig. 3.1 The neutral atmosphere.

Table 3.1
a. Permanent Constituents of Air (Troposphere)

Constituent	Formula	Molecular Mass	% by Volume
Nitrogen	N_2	28.0134	78.084
Oxygen	O_2	31.9988	20.948
Argon	Ar	39.948	0.934
Neon	Ne	20.183	18.18×10^{-4}
Helium	He	4.0026	5.24×10^{-4}
Krypton	Kr	83.80	1.14×10^{-4}
Xenon	Xe	131.30	0.084×10^{-4}
Hydrogen	H	2.01594	0.5×10^{-4}
Methane	CH_4	16.04303	2×10^{-4}

Table 3.1
b. Variable Constituents of Air (Troposphere)

Constituent	Formula	Molecular Mass	% by Volume
Water Vapor	H_20	18.0160	0–7
Carbon Dioxide	CO_2	44.00995	0.01–0.1
Ozone	O_3	47.9982	0–0.01
Sulfur Dioxide	SO_2	64.064	0–0.001
Nitrogen Dioxide	NO_2	46.0055	0–0.000002

Above the mesosphere lies the *thermosphere.* Ozone is rapidly destroyed by chemical reactions above about 80 km, and molecular oxygen is rapidly broken down into atomic oxygen. The chemical composition of the thermosphere is such that it is the most absorptive layer of the atmosphere, and the temperature in the thermosphere is seen to increase due to the absorption of solar UV rays by the ambient constituents. Above the thermosphere lies the *exosphere*, where collisions between particles are so infrequent that they move on essentially ballistic trajectories subject only to the Earth's gravitational field. For the purposes of studying its effect on spacecraft, the neutral atmosphere is obviously of greatest concern at lower orbits and may be given an upper bound of 1000 km. Because gravity tends to keep the heavier molecules closest to the Earth, and because chemical reactions and solar UV at higher altitudes have a tendency to break molecular bonds, the chemical composition of the atmosphere changes as a function of altitude. N_2 is the dominant molecule in the troposphere and stratosphere, while atomic oxygen takes predominance at 175 km. Atomic oxygen yields to helium at about 650 km, and helium in turn is replaced by hydrogen at about 2500 km.

3.2 Basic Atmospheric Physics

3.2.1 Elementary Kinetic Theory

Consider a sample of gas containing N molecules, each characterized by the same molecular mass m that is confined to a closed container of volume V and temperature T. It can be assumed that the volume of the container is large in comparison to the size of the molecules, and that between collisions the molecules move in straight lines with constant velocity. If the molecules have a uniform velocity distribution, the number of molecules in any given volume element is given by $dN = n\ dV$, where n is the number density of molecules in the container. If a single molecule of velocity v impacts a surface at an angle θ and is scattered elastically (fig. 3.2), the magnitude of the change in momentum of the molecule, Δp, is given by

$$\Delta p = mv\cos\theta - (-mv\cos\theta) = 2mv\cos\theta. \tag{3.1}$$

Each molecular collision will result in the transfer of momentum to the walls of the container. In a given time dt some amount of momentum dp will be transferred to a unit area dA. The result is a force $dF = dp/dt$ spread over a unit area dA, which is a pressure on the walls of the vessel.

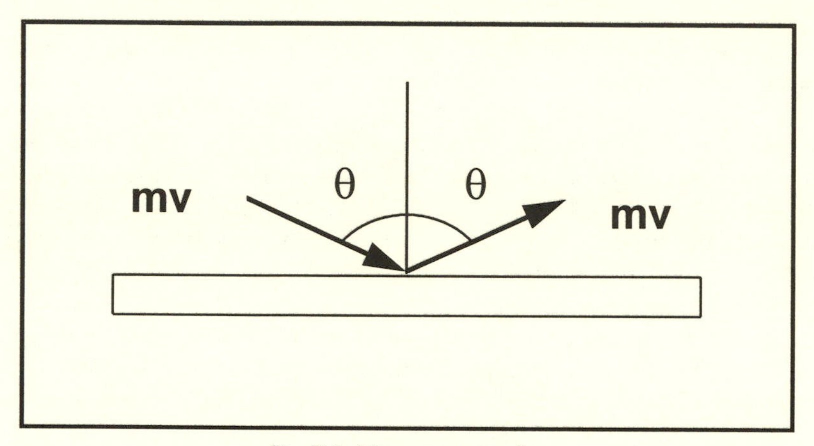

Fig. 3.2 Momentum transfer.

What is the total change in momentum due to all collisions taking place per unit area per unit time? To answer this question we must first ask, how many molecules traveling in the direction θ, ϕ with speed v strike the area in time dt? Consider those molecules contained by the volume element θ and θ + $d\theta$, ϕ and ϕ + $d\phi$, with velocity v to $v + dv$. Envision a cylinder of length vdt inclined at an angle θ to the surface element dA (fig. 3.3). All molecules satisfying the constraints on θ, ϕ, and v will strike the surface area dA in time dt. The total number of molecules in the cylinder is $dN = n\,dV = n\,v\cos\theta\,dt\,dA$. If we let dn/dv represent the number of molecules with speed between v and $v + dv$, the number of molecules with the required speed is given by

$$\frac{dn}{dv} v\cos\theta\,dt\,dA. \tag{3.2}$$

It is seen that the number of molecules having both the required speed and direction is

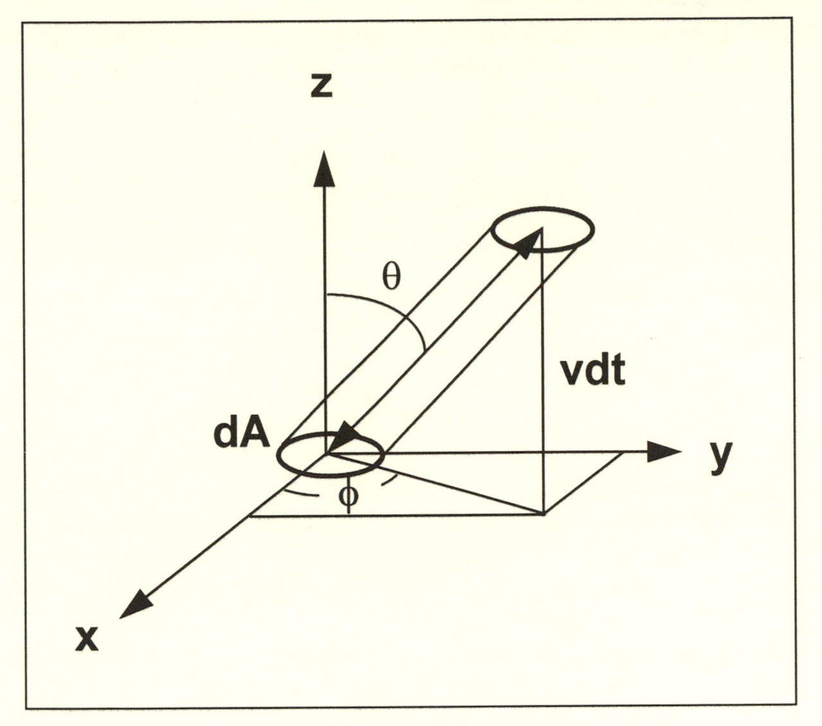

Fig. 3.3 Volume element geometry.

$$\frac{dn}{dv} v\cos\theta \, dt dA \frac{d\omega}{4\pi} \tag{3.3}$$

where $d\omega = \sin\theta \, d\theta \, d\phi$ is the solid angle that the molecules come from and the ratio $d\omega/4\pi$ is the fraction of molecules with speed v coming from this direction. Therefore, the change in momentum due to all collisions, from all directions, per unit area per unit time is given by

$$\int_0^{2\pi}\int_0^{\pi/2} (2mv\cos\theta)\left(\frac{dn}{dv} v\cos\theta \, \frac{1}{4\pi}\right)\sin\theta \, d\theta \, d\phi, \tag{3.4}$$

which reduces to

$$\frac{1}{3} mv^2 \frac{dn}{dv}. \tag{3.5}$$

The total pressure exerted on the walls of the vessel is found by integrating over all molecules, or equivalently, over the entire distribution of molecular velocities, and is given by

$$P = \frac{dF}{dA} = \frac{1}{3}m\left(\int_0^n v^2 dn\right) = \frac{1}{3}m\left(\int_0^n v^2 \frac{dn}{dv}dv\right). \tag{3.6}$$

We define the average value of the square of the velocity by

$$\bar{v}^2 = \frac{1}{n}\int_0^n v^2 \frac{dn}{dv}dv, \tag{3.7}$$

and the expression for pressure reduces to

$$P = \frac{1}{3}mn\bar{v}^2. \tag{3.8}$$

Combining this expression with the ideal gas law,

$$PV = NkT \tag{3.9}$$

and the definition $nV = N$, it can be shown that the translational kinetic energy of the molecules is related the temperature of the gas by

$$\frac{1}{2}m\bar{v}^2 = \frac{3}{2}kT. \tag{3.10}$$

When dealing with rarefied gases, such as those encountered at higher orbits, it is important to bear in mind that the concept of temperature is understood to be that expressed by equation 3.10. That is, temperature is a measure of the thermal velocity of the various atmospheric constituents. Even though the temperature of the atmosphere is on the order of 1000 K at the higher altitudes, the amount of heat transferred to an object immersed in this environment is quite low, due to the low thermal conductivity associated with the rarefied atmosphere.

The preceding discussion is based on the presumption that the molecular velocities could be described by a uniform velocity distribution. This is indeed the case, and it can be shown (exercise 1) that the number of particles per m^3 with velocity between v and $v + dv$ is given by

$$\frac{dn}{dv_i} = n\left(\frac{m}{2\pi kT}\right)^{1/2} \exp\left(-\frac{mv_i^2}{2kT}\right) \tag{3.11}$$

in one dimension, or

$$\frac{dn}{dv} = 4\pi n\left(\frac{m}{2\pi kT}\right)^{3/2} v^2 \exp\left(-\frac{mv^2}{2kT}\right) \tag{3.12}$$

in three dimensions. The probability that a particle has a velocity between v and $v + dv$ is

$$f(v)dv = \frac{1}{n}\frac{dn}{dv}dv = 4\pi\left(\frac{m}{2\pi kT}\right)^{3/2} v^2 \exp\left(-\frac{mv^2}{2kT}\right)dv. \tag{3.13}$$

This is the well-known Maxwell-Boltzmann velocity distribution function. It can also be shown (exercise 2) that the root mean square velocity is $v_{rms} = (3kT/m)^{1/2}$, while the average velocity is $v_{avg} = (8kT/\pi m)^{1/2}$. As shown in figure 3.4, a sizable fraction of the molecules in a gas may have velocities significantly above the mean thermal velocity.[4-6]

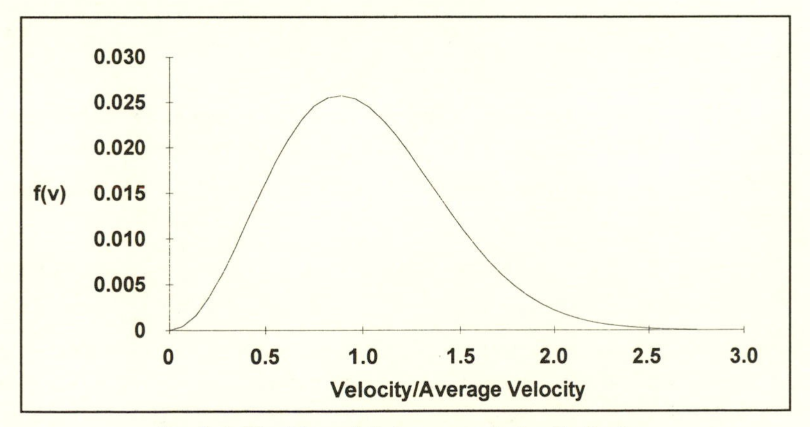

Fig. 3.4 The Maxwell-Boltzmann velocity distribution.

3.2.2 Hydrostatic Equilibrium

Consider a unit volume of gas of cross-sectional area dh (fig. 3.5). The change in pressure across the unit volume is

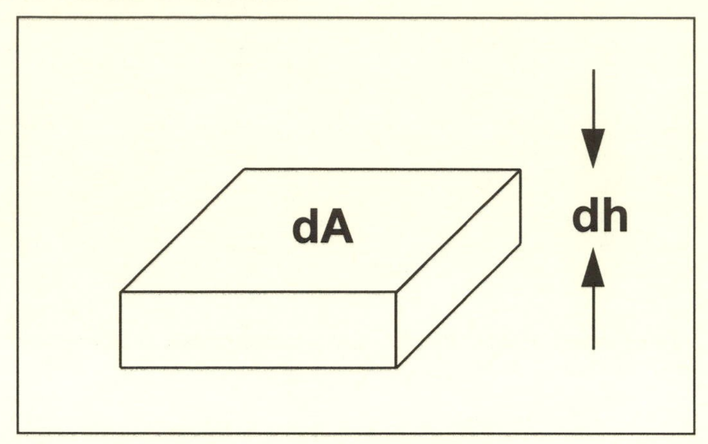

Fig. 3.5 A unit volume element.

$$dP = d\left(\frac{F}{A}\right) = \frac{dF}{A} = \frac{1}{A}d(mg) = \frac{g}{A}d(\rho Ah) = \rho gdh. \qquad (3.14)$$

Combining this equation with the ideal gas law (equation 3.9) shows that the relationship between pressure and altitude is given by

$$\frac{1}{P}dP = -\frac{mg}{kT}dh. \qquad (3.15)$$

The expression (kT/mg) is given a special symbol, H, and is called the *scale height*. Integrating equation 3.15 gives the relation between pressure and altitude, which is

$$P(h) = P_0 \exp\left(\frac{h_0 - h}{H}\right). \qquad (3.16)$$

Consequently, a general feature of atmospheric pressure is that it decreases exponentially as altitude increases. As illustrated in figure 3.6, this is a good approximation up to 100 km. However, because all of the atmospheric constituents will be in thermal equilibrium, the lighter gas molecules will have larger velocities. As a result, they are more able to resist the pull of the Earth's gravitational field and escape to higher altitudes. Similarly, due to differences in atmospheric chemistry and temperature in the different atmospheric regions, the relative abundance of species is a strong function of altitude, (fig. 3.7). As a result, the scale height is not a constant, but also varies with altitude.

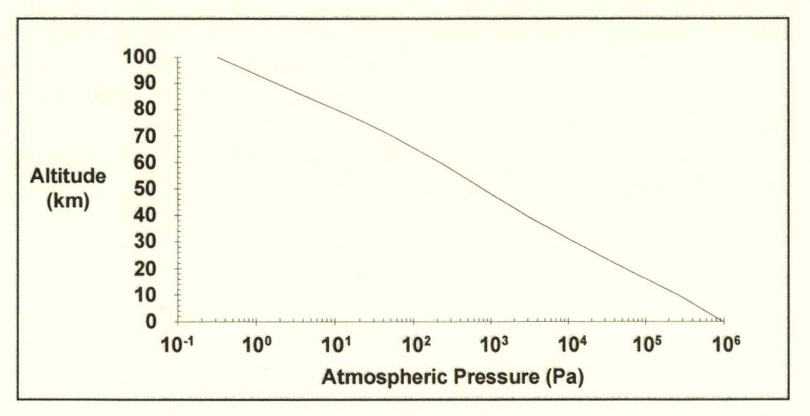

Fig. 3.6 Neutral atmospheric pressure below 100 km.

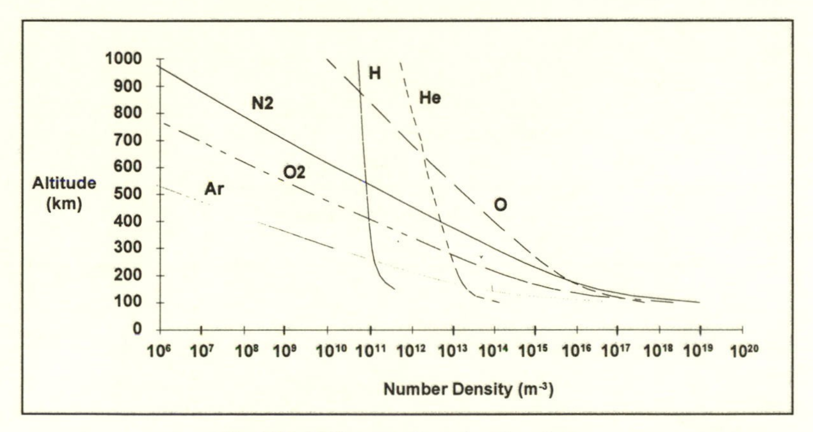

Fig. 3.7 LEO neutral species abundance.

3.2.3 Neutral Atmospheric Models

Rather than rely purely on theoretical predictions to obtain atmospheric properties, it is simpler, and usually more reliable, to utilize one of several standard models of the atmosphere. The American National Standard Guide to Reference and Standard Atmosphere Models lists ten global models of the atmosphere, with each variation having its own list of pros and cons, depending on the nature of the phenomena that the user wishes to investigate. Mass density and temperature from two of the more popular atmospheric models, the US Standard Atmosphere 1976 and the Mass Spectrometer Incoherent Scatter (MSIS-86) model,[7] are presented in figures 3.8 and 3.9, respectively. (Fig. 3.6 and 3.7 are based on the US Standard Atmosphere 1976.) The US Standard Atmosphere 1976 utilizes daily average values, based on average solar conditions. Fluctuations due to day/night variations, season, or latitude are not shown. Users interested in incorporating these variations in their calculations may wish to investigate the Mass Spectrometer Incoherent Scatter (MSIS) model. The MSIS model is a computer code that is based on both in situ observations from spacecraft and remote sensing observations from incoherent scatter radar sites. A variety of atmospheric models are available through NASA's COSMIC software library.

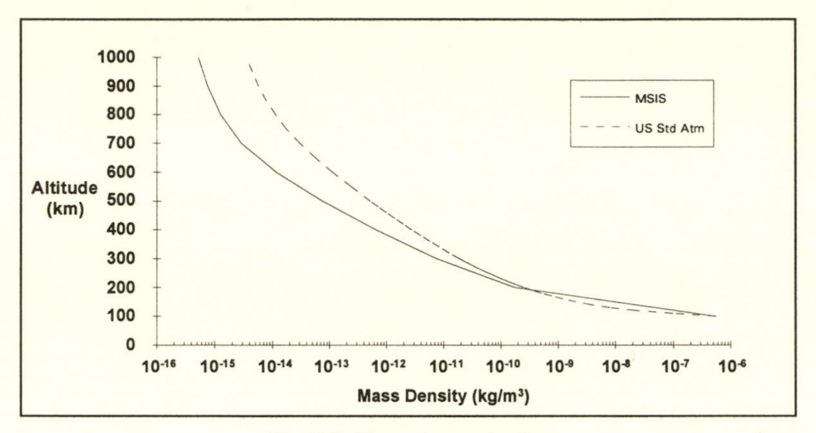

Fig. 3.8 Neutral mass density.

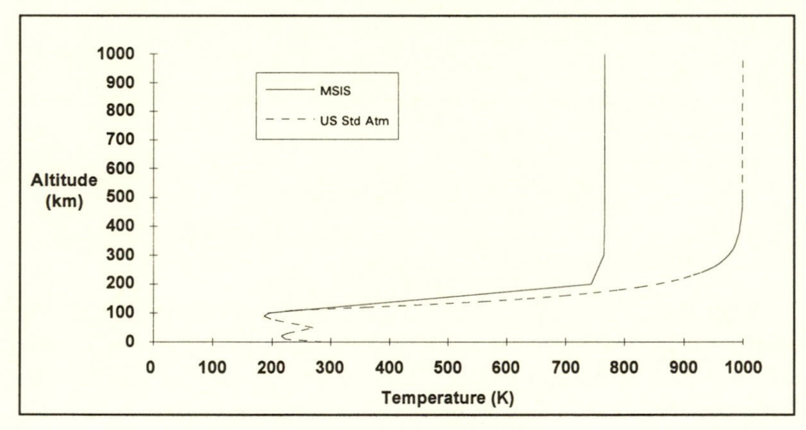

Fig. 3.9 Neutral temperature.

3.2.4 Additional Concerns

The discussion to this point has overlooked a number of issues that are of great interest in the field of aeronomy. For example, questions of energy flow into and out of the various regions have not been addressed, nor has the subject of atmospheric winds, tides, or waves. While these issues are of critical importance to obtain insight into the various atmospheric processes, they in general do not bear a direct relation to the spacecraft designer. Consequently, they will not be discussed further here. One issue that cannot be overlooked, however, is the question of how the atmosphere responds to slight variations in solar output associated with the 11-year solar cycle.[8] As can be seen from equation 3.9, heating a gas at a constant pressure will cause its volume to increase. Similarly, when the Sun deposits more energy into the Earth's atmosphere during solar maximum, the result is that the atmosphere expands and density at the higher altitudes increases. Figures 3.10 and 3.11 illustrate the general relationship between atmospheric density, temperature and the solar cycle.

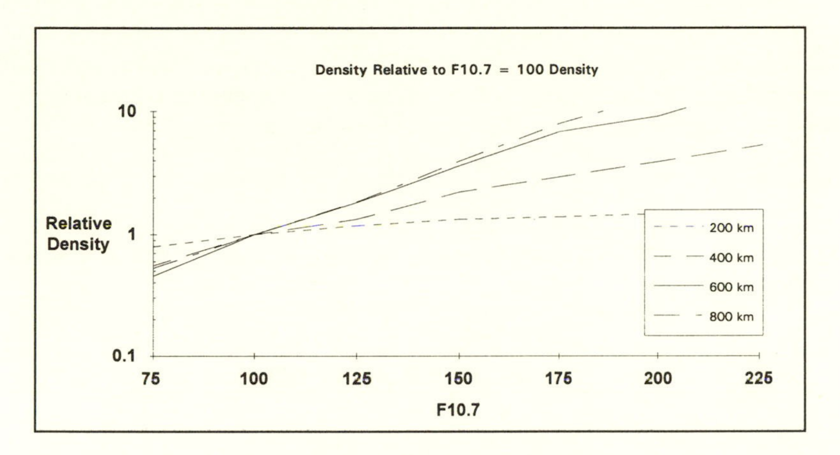

Fig. 3.10 Neutral density versus F10.7.

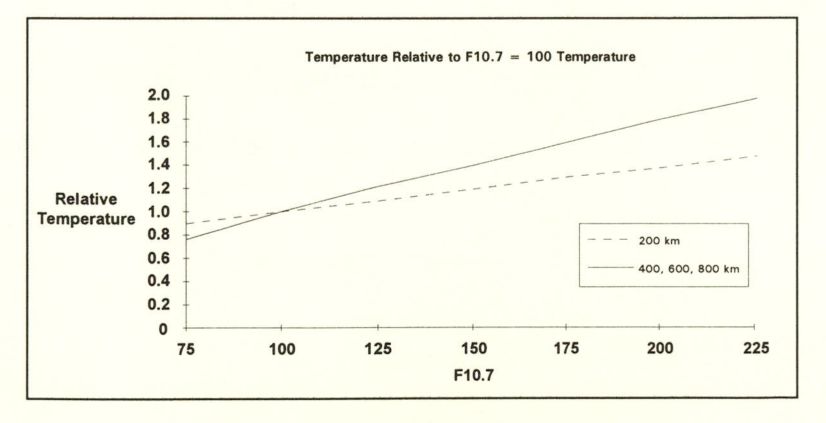

Fig. 3.11 Neutral temperature versus F10.7.

As shown, deviations from the average values are greatest for the highest altitudes. Spacecraft missions that take place at times other than average solar conditions, and/or are of long enough duration to see an appreciable change in solar activity, will need to take these variations into account in order to predict atmospheric behavior over the life of the mission.

3.3 Neutral Environment Effects

3.3.1 Mechanical Interactions

3.3.1.1 Aerodynamic Drag

Consider the interaction between an object and a collisionless flow of neutral particles of mass density ρ which has a flow velocity v with respect to the object. The particles comprising the fluid impact the object with momentum p_i and are observed to recoil from the object with momentum p_r. The amount of momentum that is transferred to the object is

$$\Delta p = p_i + p_r. \tag{3.17}$$

Because the initial momentum is usually known, equation 3.17 is commonly written in the form

$$\Delta p = p_i\left(1 + \frac{p_r}{p_i}\right) = p_i(1 + f(\theta)), \tag{3.18}$$

where the angle θ is the angle of incidence, measured with respect to the normal to the surface. In time Δt, the total amount of mass that strikes an object of cross-sectional area A is given by

$$m = \rho A v \Delta t. \tag{3.19}$$

It follows that a surface element dA, inclined at an angle θ to the fluid flow, will experience a drag force that is given by the relation

$$dF = \frac{\Delta p}{\Delta t} = \rho v^2 \left(1 + f(\theta)\right) dA. \tag{3.20}$$

By convention, we define

$$dF = \frac{1}{2}\rho v^2 C_d dA \tag{3.21}$$

where

$$C_d = 2(1 + f(\theta)) \tag{3.22}$$

is the dimensionless drag coefficient. (Note that some authors utilize the symbol $\beta = m/AC_d$ to describe the aerodynamic properties of a body.) In general, it is difficult to predict the function $f(\theta)$ theoretically due to numerous uncertainties about the nature of the individual atomic collisions. A particle may scatter elastically from a surface, or may adhere to the surface long enough to establish thermal equilibrium, before scattering randomly. The options are referred to as *specular* or *diffuse reflection* (fig. 3.12). The factors that determine the nature of a particular interaction include such variables as surface material, surface temperature, impacting species, etc. The majority of collisions are at least partially diffuse in nature and, consequently, experimental determinations of C_d are usually needed. An average value based on the results of previous investigations is shown in figure 3.13.[9-11]

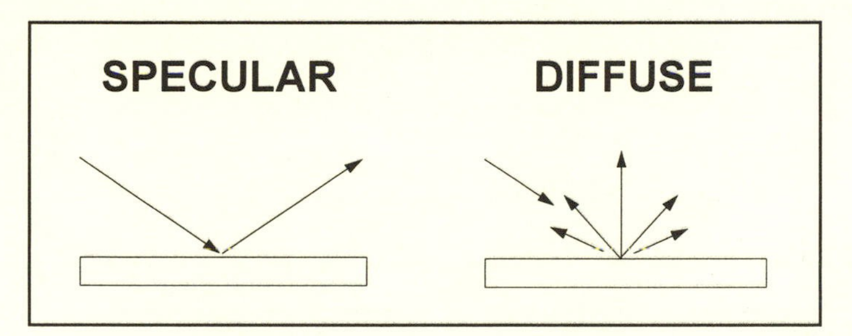

Fig. 3.12 Specular and diffuse reflection.

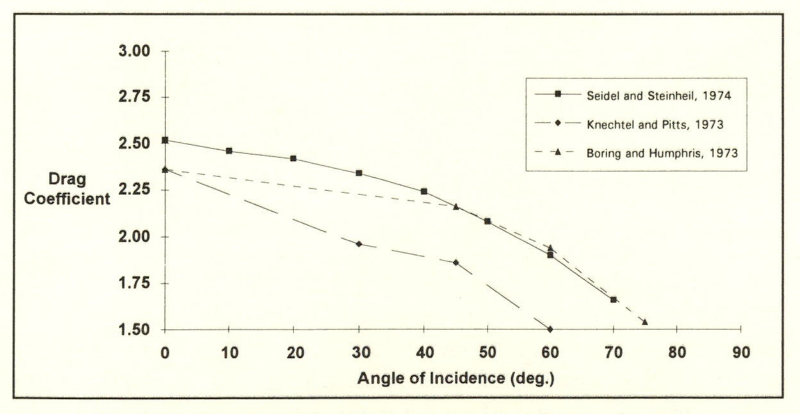

Fig. 3.13 Experimentally determined values of C_d for a flat plate. (©1985 AIAA—reprinted with permission)

To obtain an expression for the total drag force on an object, it is necessary to integrate equation 3.21 over the entire surface

$$F = \frac{1}{2}\rho v^2 \oint C_d dA. \tag{3.23}$$

Equation 3.23 can be generalized to apply to any shapes: plates, cones, spheres, etc. The final result is given by the well-known drag force equation

$$F = \frac{1}{2}\rho v^2 C_d A, \tag{3.24}$$

where A is the area of the object normal to the flow and C_d is the drag coefficient for the entire object. The only constraint is that the impact angle must be low enough to prohibit self-shadowing of the surface (which may arise due to thermal motions of the particles in the flow). Table 3.2 lists examples of C_d for a variety of shapes.[12-14] For most spacecraft C_d is typically on the order of 2.20, though values as low as 1.9 and as high as 2.6 have been reported.[9,15,16]

Table 3.2
Drag Coefficients for Various Shapes

Shape	Drag Coefficient
Flat plate Inclined at angle $\theta < 85°$	$2(1+f(\theta))$
Cone Half angle $\phi > 5°$	$2\left(1+f\left(\frac{\pi}{2}-\phi\right)\right)$
Truncated cone Half angle $\phi > 5°$ Inner/outer radius b/a	$2(1+f(0°))\left(\frac{b}{a}\right)^2 + 2\left(1+f\left(\frac{\pi}{2}-\phi\right)\right)\left(1-\left(\frac{b}{a}\right)^2\right)$

C_d is a measure of the extent to which the incident flow is accommodated to the surface of the impacted object. It is common for many authors to define surface reflection coefficients and an energy transfer coefficient as a measure of surface accommodation.[10,16-18] The tangential and normal momentum coefficients, σ_t and σ_n, are defined by

$$\sigma_t = \frac{p_{i,t} - p_{r,t}}{p_{i,t}} \qquad \sigma_n = \frac{p_{i,n} - p_{r,n}}{p_{i,n} - p_{w,n}}, \tag{3.25}$$

where the subscripts i, and r refer to incident and reflected momentum, respectively, and the subscript w is used to refer to those particles that have established thermal equilibrium with the surface. The thermal accommodation coefficient is given by

$$\alpha = \frac{E_i - E_r}{E_i - E_w}. \tag{3.26}$$

These accommodation coefficients can be thought of as the measure of the reflected molecules that are re-emitted diffusively.

In addition to the drag force on surfaces oriented normal to the flow direction, there is also the possibility for drag on laterally oriented surfaces.[12] This may be significant for microgravity missions, spacecraft sensitive to aerodynamic torques, or missions where fuel consumption is the critical life-limiting factor. The thermal velocity of a typical oxygen atom in LEO is on the order of 1 km/s. This is just a fraction of the orbital velocity of ~ 8 km/s, but still large enough for many atoms to impact lateral surfaces of the vehicle, as shown in figure 3.14. Each of these collisions may in turn transfer momentum to the vehicle.

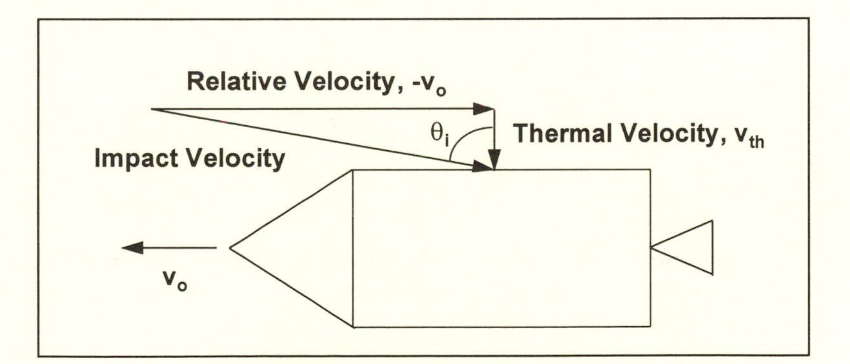

Fig. 3.14 Lateral surface drag.

We define the lateral drag force by

$$F_{ls} = \frac{1}{2}\rho v^2 C_{d,ls} A_{ls}, \tag{3.27}$$

where A_{ls} is the lateral surface area,

$$C_{d,ls} = \left(1 + f(\theta_i)\right)\cot\theta_i \tag{3.28}$$

and

$$\theta_i = \tan^{-1}\left(\frac{v}{v_{th}}\right). \tag{3.29}$$

The total drag coefficient is more appropriately defined as

$$C_D = C_d + C_{d,ls}\frac{A_{ls}}{A}. \tag{3.30}$$

Utilizing this expression in equation 3.23 will give a drag force that includes contributions from both normal and lateral surfaces. As a rule of thumb, lateral drag forces are small in comparison to normal forces unless large lateral surfaces (such as solar panels) are present. The drag force will eventually lead to orbital decay and reentry (fig. 3.15).[8]

Example 3.1

Calculate the normal drag coefficient for the configuration shown below.

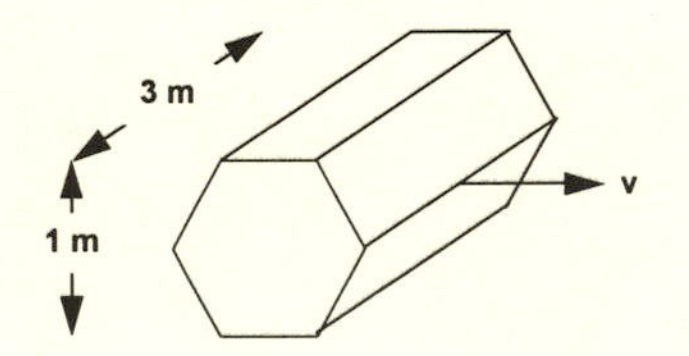

By inspection it can be seen that the angle between the velocity vector and the normal to the surface is 30°. Consequently, from figure 3.13 it can be seen that C_d is ~ 2.20.

Consider the example of a spacecraft of dry mass m and fuel mass m_f that is moving at velocity v_0 subject to a drag force F. In order to counter the drag force, fuel is expelled from the spacecraft at a relative velocity v'. Since mass is conserved, the change in the velocity of the spacecraft is given by

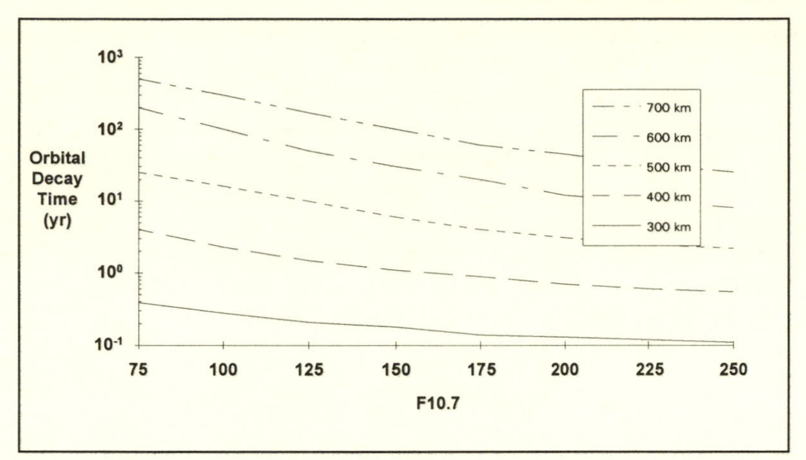

Fig. 3.15 Orbital decay time.

$$(m+m_f)\frac{\Delta v_o}{\Delta t} = -F + m_f \frac{v'}{\Delta t}. \tag{3.31}$$

The orbit will be maintained if $\Delta v_o = 0$. This constraint reduces to

$$F = m_f \frac{v'}{\Delta t}. \tag{3.32}$$

In general, chemical or electrical propulsion systems operate by expelling fuel, or a working gas, at a fixed constant velocity, v'. Consequently, the problem of maintaining orbit reduces to (i) choosing the method of propulsion (i.e., value of v'), and (ii) estimating the amount of fuel needed to maintain orbit. Rather than speak of exhaust velocities, it is common for propulsion designers to use the term *specific impulse*, I_{sp}, which is defined by

$$I_{sp} = \frac{Thrust}{\frac{\Delta m}{\Delta t} g} \tag{3.33}$$

and related to the exhaust velocity, v', by the relation

$$v' = I_{sp} g. \tag{3.34}$$

Examples of I_{sp} and v' for typical spacecraft propulsion systems are shown intable 3.3. Although the amount of fuel needed to counter drag is a significant factor in spacecraft design, the choice of a propulsion system must involve many other factors. Bipropellant systems have a higher specific impulse than monopropellant ones, but require additional fuel tanks, valves, and tubing to operate. Similarly, electrical propulsion systems require the spacecraft to provide continuous power for their operation. All of these factors will add weight and complexity to the vehicle and must be part of an overall system level trade to identify the most appropriate solution.

Table 3.3
Specific Impulse for Various Propulsion Methods

Propellant	Vacuum I_{sp} (s)	Thrust (N)
Cold Gas		
N_2, NH_3, Freon, He	50–75	0.05–200
Liquid, Monopropellant		
H_2O_2, N_2H_4	150–225	0.05–0.5
Liquid, Bipropellant		
N_2O_4 and MMH	300–340	$5–5 \times 10^6$
Solid		
Organic Polymers	280–300	$50–5 \times 10^6$
Electrothermal		
Resistojet, Arcjet	150–1500	0.005–5
Electrostatic		
Colloid, Ion	1200–6000	5×10^{-6}–0.05
Electromagnetic		
Pulsed Plasma/Inductive, MPD	1500–2500	5×10^{-6}–200

Example 3.2

Calculate the drag makeup fuel needed for the spacecraft in example 3.1 to maintain a 200 km circular orbit for one year. Assume average solar cycle conditions and monopropellant fuel.

From figure 3.8, the atmospheric mass density at 200 km is ~ 4 × 10^{-10} kg/m^3. The orbital velocity is ~ 8 km/s and the cross sectional area of the spacecraft is (3 m)(1 m) = 3 m^2. The drag force on the spacecraft is

$$F = (0.5)(4 \times 10^{-10}\ \text{kg/m}^3)(8000\ \text{m/s})^2(2.20)(3\ \text{m}^2) \sim 0.01\ \text{N}.$$

For monopropellant fuel, I_{sp} ~ 200 s so $v' = I_{sp}g$ = (200 s)(9.8 m/s^2) = 1960 m/s. The mass of fuel needed for one year is

$$m_f = F\Delta t/v' = (0.01\ \text{N})(3.15 \times 10^7\ \text{s})/(1960\ \text{m/s}) = 160.7\ \text{kg}.$$

3.3.1.2 Physical Sputtering

Neutral molecules which impact spacecraft in circular orbits do so with nonnegligible energies as shown in table 3.4. There is the possibility for each of these collisions to sever the chemical bond of a surface atom to that of its neighbors, provided that the energy of the bond is less than the impact energy. This process is called *sputtering*.

Table 3.4
LEO Impact Energies—Circular Orbits

Altitude	Velocity	Species Energy (eV/particle)					
(km)	(km/s)	H	He	O	N_2	O_2	Ar
200	7.8	0.3	1.3	5.0	8.8	10.1	12.6
400	7.7	0.3	1.2	4.9	8.6	9.8	12.2
600	7.6	0.3	1.2	4.7	8.3	9.5	11.8
800	7.4	0.3	1.1	4.5	7.9	9.0	11.2

In order for sputtering to occur, the impact energy must first exceed the minimum energy needed to bind a surface atom to its neighbors. Consequently, sputtering is characterized by an energy threshold that is approximated by

$$E_{th} = 8U\left(\frac{m_t}{m_i}\right)^{-1/3}, \qquad \text{for } m_t/m_i < 3$$
$$E_{th} = U[\gamma(1-\gamma)], \qquad \text{for } m_t/m_i > 3, \qquad (3.35)$$

where

$$\gamma = \frac{4m_t m_i}{(m_t + m_i)^2}. \tag{3.36}$$

U is defined to be the binding energy of a surface atom (similar to the activation energy discussed in the previous chapter), m_i is the incident particle mass, and m_t is the target-atom mass.[19] In general, sputtering thresholds for most materials are above the average impact energy (table 3.5).

Table 3.5
Sputtering Thresholds

Target	Bombarding Gas Threshold (eV)					
Material	O	O_2	N_2	Ar	He	H
Ag	12	14	13	17	25	83
Al	23	29	27	31	14	28
Au	19	15	15	15	53	192
C	65	82	79	88	40	36
Cu	15	22	21	24	20	60
Fe	20	28	27	31	23	66
Ni	20	29	27	31	24	72
Si	31	39	37	42	18	40

Source: From *Planet. Space Sci.*, 36, R. R. Laher and L. R. Megill, "Ablation of Materials in the Low Earth Orbital Environment," pp. 1497–1508, © 1988. (Reprinted with kind permission of Pergamon Press Ltd, Headington Hill Hall, Oxford OX3 OBW, UK.)

As the result of the high-energy tail in the Maxwell-Boltzmann velocity distribution some ambient neutrals can impact at energies above the threshold, but this is rare.[20] As we will see in the next chapter, spacecraft charging can greatly increase the impact energy of charged molecules, leading to sputtering of surface material during severe charging conditions. The ratio of sputtered surface molecules per ambient collision is termed the sputtering yield. For low energies, the sputtering yield is approximated by the semi-emperical formula

$$Y^i(E) = Q^i \left(\frac{E}{E^i_{th}} \right)^{0.25} \left(1 - \frac{E^i_{th}}{E} \right)^{3.5}, \tag{3.37}$$

where $Y(E)$ is the sputtering yield, Q is a normalization factor, and E is the impact energy.[19] The superscript i is used to emphasize that these quantities are dependent on the bombarding species as well as the target material. Figure 3.16 illustrates the general relationship between yield and impact energy, while table 3.6 lists typical values for sputtering yields at 100 eV impact energy.[20]

The total flux of material ablated from the surface, ϕ_s, is given by

$$\phi_s = \sum_i \int_{E_{th}^i}^{\infty} Y^i(E)\phi_i(E)dE, \tag{3.38}$$

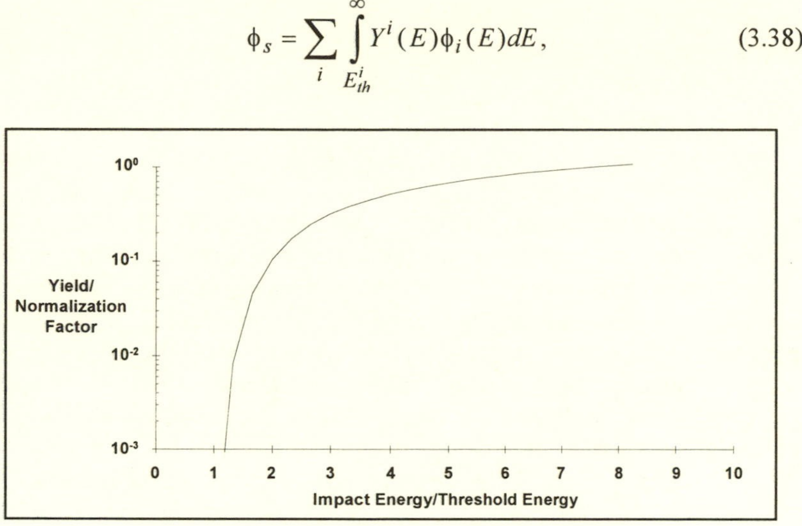

Fig. 3.16 Sputtering yield versus impact energy.

Table 3.6
Sputtering Yields at 100 eV Impact Energy

Target Material	Sputtering Yields (atoms/particle)					
	O	O_2	N_2	Ar	He	H
Ag	0.265	0.498	0.438	0.610	0.030	-
Al	0.026	0.076	0.060	0.110	0.020	0.010
Au	0.154	0.266	0.244	0.310	-	-
C	-	-	-	-	0.008	0.008
Cu	0.385	0.530	0.499	0.600	0.053	-
Fe	0.069	0.153	0.129	0.200	0.028	-
Ni	0.120	0.247	0.239	0.270	0.029	-
Si	0.029	0.054	0.046	0.070	0.023	0.002

Source: From *Planet. Space Sci.*, 36, R. R. Laher and L. R. Megill, "Ablation of Materials in the Low Earth Orbital Environment," pp. 1497–1508, © 1988. (Reprinted with kind permission of Pergamon Press Ltd, Headington Hill Hall, Oxford OX3 OBW, UK.)

where ϕ_s is in units of sputtered particle flux, $\phi_i(E)$ is the flux of species i incident particles with energies between E and dE, and the sum is carried out over all atmospheric species i. In the absence of spacecraft charging (which will be discussed in the next chapter) ablation rates of most materials are acceptably small.[20] It is only on very long duration missions, such as the 30-year lifetime of the space station, that sputtering of materials may be a significant factor in determining material lifetimes.

3.3.2 Chemical Interactions

3.3.2.1 Atomic Oxygen Attack

As was previously shown, between roughly 100 and 650 km altitude the main atmospheric constituent is atomic oxygen (AO). In addition to the mechanical interactions just discussed, spacecraft in LEO are subjected to a variety of chemical interactions that arise due to the reactive nature of AO. AO is known to interact with a wide variety of materials, leading to oxidation or erosion and general degradation of materials properties.[21-23] Depending on the nature of the material in question, the effects of the AO environment may be very damaging, very subtle, or even beneficial. Most combustible materials are heavily etched, and some coatings, such as silver and osmium, are seriously degraded or removed as volatile oxides. Because spacecraft materials are, in general, chosen for their thermal properties and, at the same time, are kept as thin as possible to minimize launch weight, AO degradation may have potentially serious consequences if the mass loss leads to a change in thermal characteristics.

Consider the erosion of material from a surface due to a flux of AO atoms. If the flux acts for a time dt, the mass loss from a surface area dA is described by the equation

$$dm = \rho \mathrm{RE} \phi \, dA \, dt , \tag{3.39}$$

where ρ (g cm^{-3}) is the density of the material, ϕ (AO atom cm^{-2} s^{-1}) is the AO flux, which is the product of number density and orbital velocity, and RE (cm^{3} AO atom^{-1}) is an experimentally determined constant of proportionality called the *reaction efficiency*. Reaction efficiency is useful in this context because it simplifies the relationship between mass loss and AO flux. Because the physical processes involved with mass loss are time dependent and fairly complicated, it is important to bear in mind that RE may be a function of surface temperature, AO flux, coincident solar UV energy, surface contamination, and other parameters. Related to RE is the reaction efficiency ratio (RER), which is the ratio of the RE of a given material to that of a control sample, most often Kapton® (Dupont Corp.). Equation 3.39 may be rewritten in terms of the rate of change in the thickness of the material, dx/dt, as

$$\frac{dx}{dt} = \mathrm{RE}\phi . \qquad (3.40)$$

This issue of AO degradation is of greatest concern for long-term missions, and several experiments have been flown in preparation for the 30-year life of the space station.[24] (The Long Duration Exposure Facility [LDEF], which spent 5 years and 10 months in LEO and returned a wealth of data on materials degradation, is discussed separately in appendix 4.) Experiments on STS-8 and STS-41G have recorded a great deal of in situ data to test material erosion rates, and the effects of protective coatings.[25-27] On STS-8 more than 300 individual samples were exposed to ram conditions experiencing a total AO fluence of 3.5×10^{20} atoms/cm^2. The STS-8 experiment included samples of polymer-matrix and metal-matrix components; white-paint coatings; black-paint coatings; second-surface mirror coatings; chromic acid anodized coatings; and sputter-deposited coatings. The STS-41G experiment was designed to measure the effect of AO on polymeric-based spacecraft materials. Table 3.7 illustrates the relative surface recession rates for a variety of materials subjected to the in situ AO environment. Figure 3.17 illustrates the effect of the AO environment on the surface of an epoxy composite; the mass loss is quite evident.

The polyimide film Kapton® is used in the shuttle payload bay and on satellites as a component of thermal control blankets because of its high temperature and UV stability, and toughness. Consequently, it is a good material to single out as an example.

Example 3.3

Calculate the rate at which Kapton would be eroded from a spacecraft in a 200 km circular orbit.

From table 3.7, the reaction efficiency of Kapton is 3.04×10^{-24} cm^3/atom. The number density of AO, from figure 3.7, is 5×10^{15} m^{-3}. Therefore the erosion rate is

$$dx/dt = \mathrm{RE}\ \phi = (3.04 \times 10^{-24}\ \mathrm{cm^3/atom})(5 \times 10^{15}\ \mathrm{m^{-3}})(1\ \mathrm{m}/100\ \mathrm{cm})^3 \times (8 \times 10^5\ \mathrm{cm/s})$$

$$dx/dt = 1.2 \times 10^{-8}\ \mathrm{cm/s} = 0.38\ \mathrm{mm/yr} = 150\ \mathrm{mils/yr}.$$

Kapton® can be protected by thin films (< 1000 angstroms) of Al_2O_3 and SiO_2, as well as a mixture of predominantly SiO_2 with a small amount of

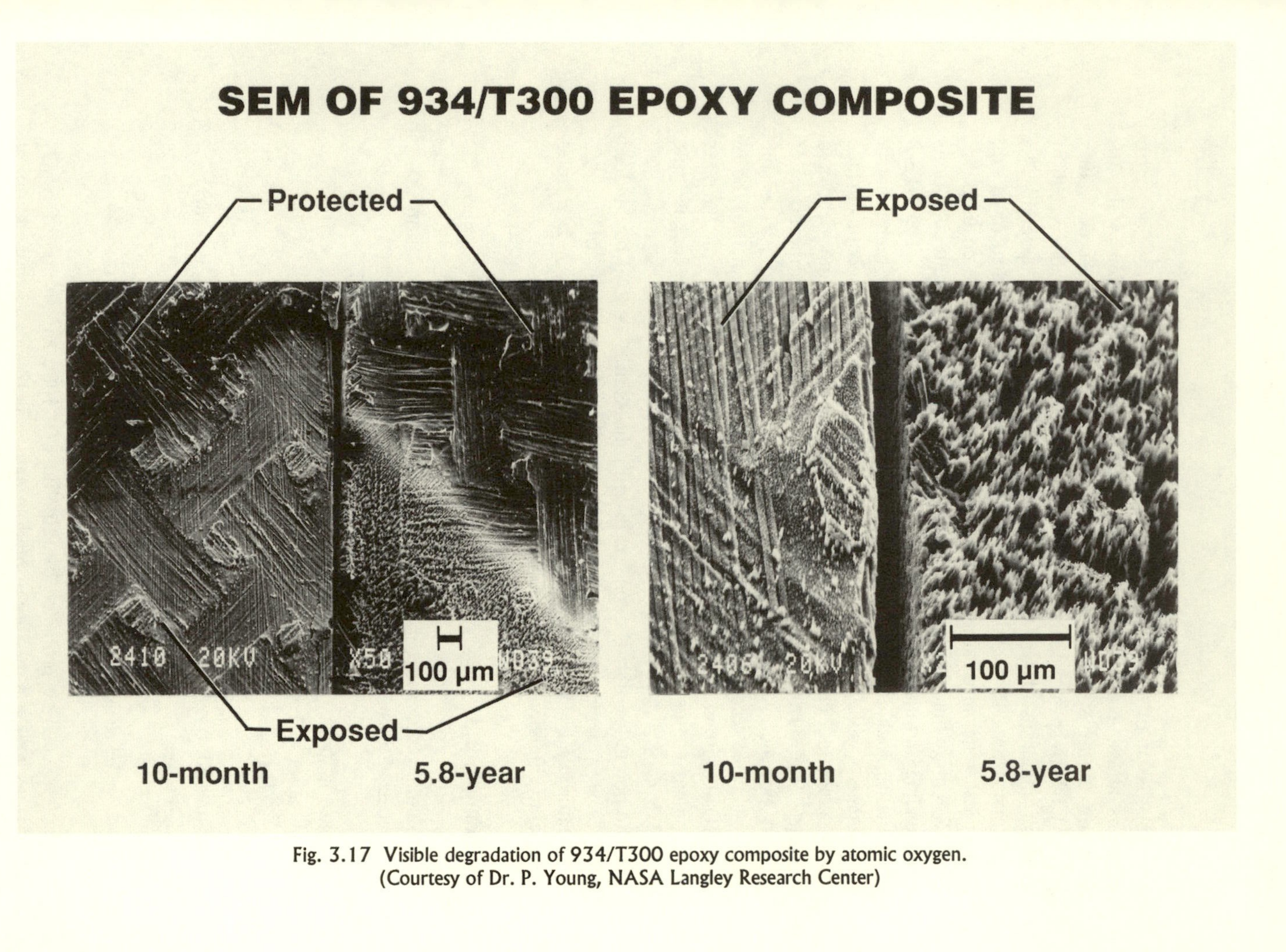

Fig. 3.17 Visible degradation of 934/T300 epoxy composite by atomic oxygen. (Courtesy of Dr. P. Young, NASA Langley Research Center)

polytetrafluoroethylene (PTFE) (table 3.8).[28,29] The latter resulted in a mass loss rate that was 0.2 percent that of unprotected Kapton®. AO coatings must typically be thin in order to maintain the thermal properties of the materials they protect, and it has been verified in ground tests that the presence of the thin coatings did not significantly alter the optical properties of the Kapton® between 0.33 and 2.2 μm wavelength.[27]

Table 3.7
Atomic Oxygen Reaction Efficiency

	Reaction Efficiency ($\times 10^{-24}$ cm^3/atom)	
Material	Range	Best Value
Aluminum	-	0.00
Carbon	0.9–1.7	-
Epoxy	1.7–2.5	-
Fluoropolymers		
- FEP Kapton	-	0.03
- Kapton F	-	<0.05
- Teflon, FEP	-	<0.05
- Teflon	0.03–0.50	-
Gold	-	0.0
Indium Tin Oxide	-	0.002
Mylar	1.5–3.9	-
Paint		
- S13GLO	-	0.0
- YB71	-	0.0
- Z276	-	0.85
- Z302	-	4.50
- Z306	-	0.85
- Z853	-	0.75
Polyimide		
- Kapton	1.4–2.5	-
- Kapton H	-	3.04
Silicones		
- RTV560	-	0.443
- RTV670	-	0.0
Silver	-	10.5
Tedlar		
- Clear	1.3–3.2	-
- White	0.05–0.6	-

There are concerns for other materials in addition to Kapton®. Metals such as silver or aluminum, which are often used as the reflective surfaces of mirrors, will oxidize upon exposure to the AO environment. Consequently, they too may need to be protected by coatings (table 3.9).[30,31] There is also the possibility for erosion of solar array interconnects. As shown in figure 3.18, a typical 2 cm × 4 cm solar cell will be connected to its neighbors with 2-40 × 30 mil silver wires that are about 1 mil in thickness. If no protective measures are taken, the interconnects will start to erode upon exposure to the LEO environment. After prolonged exposure, the interconnects may literally be worn away and the solar array will fail. Consequently, coating the interconnects may be necessary for long missions in especially low orbits.

Table 3.8
Mass Loss Rates for Kapton®

Protective Coating	Thickness (angstroms)	Mass Loss (mg)	Mass Loss per incident O atom (g/atom)
None	0	5020 +/– 9.9	4.3×10^{-24}
Al_2O_3	700	567 +/– 5.2	4.8×10^{-25}
SiO_2	650	5.9 +/– 5.2	5.0×10^{-27}
96% SiO_2 + 4% PFTE	650	10.3 +/– 5.2	8.8×10^{-27}

Table 3.9
Change in Solar Reflectance of Silver Mirror Samples

Coating	Exposure Time (hr)	Solar Reflectance		Fractional Loss of Reflectance per 1000 hr
		Start	Finish	
Ni Substrate				
Al_2O_3	75	0.911	0.911	0
SiN_4	400	0.916	0.881	0.088
Glass Substrate				
SiO_2	634	0.972	0.937	0.055
SiO_2/MgF_2	634	0.970	0.927	0.068
ITO	225	0.899	0.908	-
ITO/MgF_2	225	0.925	0.902	0.102
PFTE - SiO_2	159	0.971	0.951	0.126

Because AO degradation is chemical rather than mechanical in nature, it does not depend on a direct line of sight to induce damage. Any defect in a protective coating, which may arise due to manufacturing defects or the micrometeoroid/orbital debris environment, would allow the AO to penetrate under the surface and attack the underlying material. Similarly, AO will be able to reflect off the solar array itself and attack the back sides of solar array interconnects. In order to determine if AO is a problem, it is first necessary to compute the total flux of AO anticipated during the course of the mission. The flux can then be related to material erosion/degradation rates, which can in turn be related to subsystem degradation tolerances. If the degradation is unacceptable, coating of surfaces, or orienting sensitive surfaces away from ram, may be necessary.[32]

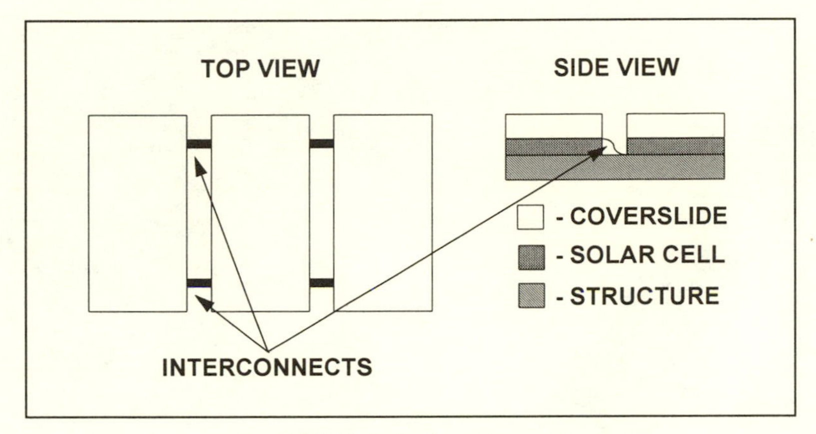

Fig. 3.18 Typical solar cell geometry.

3.3.2.2 Spacecraft Glow

Numerous spacecraft have observed optical glow emissions near exterior surfaces (fig. 3.19). Because many spacecraft are designed to carry optical instruments, there is a possibility that photometric observations could be degraded by glow from the vehicle itself. Glow has been reported on a number of unmanned spacecraft (NASA's Atmospheric Explorer C and E, Dynamics Explorer B, US Air Force Space Test Program 78-1) as well as the space shuttle.[33-37] The physical mechanism for the glow is not well understood, but it is directly related to the presence of the neutral atmosphere. The brightness of the glow, *B*, as a function of altitude, *H*, is approximated by the relation[37]

$$\log B(\text{Rayleighs}) = 7 - (0.0129)H(\text{km}), \qquad (3.41)$$

where 1 Rayleigh = $(1/4\pi) \times 10^{10}$ photons m^{-2} s^{-1} sr^{-1}. There appear to be some differences in the type of glow associated with small satellites and that associated with the space shuttle. This may be an indication that size, or relative geometry, is an important factor in controlling the nature of the glow mechanism. Similarly, some glow is associated with ram facing surfaces, while other observations indicate a glow associated with thruster firings.[38] Different mechanisms appear to be responsible for the variety of observations reported. Nevertheless, some general conclusions can be noted. First, glow increases in brightness toward the red, peaking at about 6800 angstroms. Second, glow intensity varies from material to material, with black chemglaze and Z302 glowing the brightest, polyethylene the least bright.[39] Glow from satellites, thruster firings, shuttle ram surfaces, and the shuttle's co-orbiting cloud of contaminants are all somewhat different in appearance.

Fig. 3.19 Spacecraft glow near the vertical stabilizer of the shuttle orbiter. (Photograph courtesy of Dr. Gary Swenson, Lockheed)

The main effect of glow is on remote sensing observations from optical payloads. It is tempting to suggest utilizing materials on external surfaces that are not susceptible to glow. However, experience indicates that those materials that are not susceptible to glow are susceptible to atomic oxygen attack, and vice versa. Similarly, because most materials are chosen for their thermal characteristics or stability to long-term LEO exposure, the issue of glow may make it difficult to find resolution through materials choices alone. Spacecraft flying remote sensing instrumentation in LEO may need to orient sensors into the wake or allow for degradation of image quality due to the glow phenomenon.

Example 3.4

Calculate the brightness of the glow associated with a spacecraft in a 200 km circular orbit.

From eq. 3.41, $\log B = 7 - 0.0129\, H = 7 - (0.0129)(200) = 4.42$

$$B = 10^{4.42} = 26{,}303 \text{ Rayleighs} = 2 \times 10^{13} \text{ photons m}^{-2}\text{ s}^{-1}\text{ sr}^{-1}$$

3.4 Space and Ground-Based Testing

A variety of experimental techniques have been investigated as a means to study the interaction of materials with the LEO atomic oxygen environment. Exact simulation of the LEO environment is complicated by the fact that molecular oxygen must first be dissociated into atomic oxygen and then accelerated to the 5 eV energies associated with orbital impact. Techniques for producing such a flux may utilize a high-energy laser to induce breakdown of molecular oxygen, followed by a rapid expansion of the resulting plasma,[40] or grazing-incidence collisions between ion beams and metal surfaces,[41-43] among other techniques. Alternatively, a simple low-temperature oxygen plasma reactor can be constructed from an evacuated glass cylinder wrapped by a solenoid copper coil. Oxygen is fed into the chamber with mass flow controllers, and operating pressure is maintained by using rotary vacuum pumps and appropriate regulating valves. The plasma is created, and sustained, by RF energy from the solenoid coil. In the past, the use of plasma reactors to study material degradation was questionable due to the fact that the AO collisional energy that could be obtained in ground-based simulations was substantially lower than that occurring on orbit. Although the orbital plasma environment cannot be exactly simulated, it is possible for properly calibrated plasma reactors to suitably match material RERs obtained by flight data. It is this fact that enables the low energy ground-based tests to simulate the effects of mass loss without simulating the exact LEO environment. It is important to note that in these tests there may be a variety of test conditions (reactor pressure, oxygen flow rate, voltage, current, phase of power supply, etc.) that will produce a low-energy oxygen plasma. However, only within a restricted set of reactor operating conditions will the RERs correlate well with flight data. It is important that these operating conditions be understood by the individuals conducting the tests in order to insure applicability of final results.

3.5 Design Guidelines

As shown in table 3.10 there are a variety of options available to the designer who wishes to minimize interactions with the neutral atmosphere. The magnitude of aerodynamic drag is controlled by the local atmospheric density and the shape/size of the vehicle. With rare exception, raising the operational altitude in order to minimize drag is generally prohibited due to the higher launch costs that this involves and concern over the performance of the payload. Similarly, designing an aerodynamically smooth vehicle is usually prohibited due to volume constraints imposed by the launch vehicle shroud. Consequently, drag is typically minimized by flying the vehicle in the orientation that minimizes the amount of normal surface area exposed to the flow. At lower altitudes, orienting solar panels at right angles to the ram direction is a common technique, provided that this orientation provides for enough electrical power to the vehicle. Sputtering is managed by proper choice of materials.

Atomic oxygen attack is minimized by orienting sensitive surfaces away from ram, if possible, and through the use of protective coatings. Obviously, adding AO protective coatings adds cost and complexity to the design; consequently they should only be used if problems are foreseen. In order to minimize glow it is necessary to orient remote sensing instruments toward the spacecraft wake, or away from surfaces that might glow. Also, if the option is available, and if glow is perceived to be a problem, choice of materials may minimize the glow intensity.

3.6 Summary

At 300 km the ambient atmospheric density is about ten orders of magnitude below that encountered at sea level. Relatively speaking, this is

Table 3.10
Neutral Environment Effects Design Guidelines

Materials	Choose materials that (a) are resistant to AO, (b) do not glow brightly (if optical instruments present), and (c) have high sputtering thresholds
Configuration	Aerodynamic drag may be minimized by flying the vehicle with a low cross-sectional area perpendicular to ram. Orient sensitive surfaces and optical sensors away from ram.
Coatings	Consider protective coatings for surfaces that are susceptible
Operations	If possible, fly at altitudes that minimize interactions

quite a small fraction, but in absolute terms this equates to $\sim 10^{15}$ oxygen atoms per cubic meter. A 1 m^2 spacecraft orbiting at 8 km/s would undergo about 3×10^{22} collisions with ambient atoms every hour. As the result of these collisions, the spacecraft (1) will be subjected to a drag force (causing orbital decay), (2) may experience material erosion due to sputtering, (3) may experience material degradation due to atomic oxygen attack, and (4) may generate a significant amount of visible glow. The drag force will eventually cause the vehicle to slow and reenter the atmosphere unless countered by thrusters onboard the spacecraft. This requires the spacecraft to carry fuel for this purpose, and sizing a propulsion subsystem to counter the anticipated drag force is a major aspect of spacecraft design. The sputtering process is not of concern for most typical materials exposed for lifetimes of less than a year. However, it may be significant on multiyear missions such as the space station or in the presence of spacecraft charging. Due to the reactive nature of atomic oxygen, many materials will be subjected to a variety of chemical reactions which may alter their thermal properties or reduce their thicknesses. Protecting typical spacecraft materials, such as Kapton or silvered Teflon, from the oxygen environment is an intensive area of research. Finally, it has been well documented that the neutral environment often gives rise to an optical glow near the surfaces of certain materials. This poses a concern for spacecraft flying sensitive optical payloads. Although interactions with the neutral environment are, in general, ignorable above 1000 km, spacecraft launches to LEO include about one third of all payloads and virtually all large and manned spacecraft, making these issues of significance to insure effective spacecraft design. Because each of these phenomena is directly related to atmospheric density, they are in turn dependent on the solar cycle.

3.7 Exercises

1. Define a one-dimensional normalized distribution function of a single velocity component, v_i, by

$$f(v_i) = \frac{1}{n}\frac{dn}{dv_i}.$$

For any values of v_i and v_j it is possible to define a rotation to a new set of coordinates, α and β, where $v_\alpha^2 = v^2 + v_j^2$ and $v_\beta^2 = 0$. Consequently, it follows that

$$f(v_\alpha)f(0) = f(v_i)f(v_j).$$

a. Differentiate the above expression, with respect to v_i and v_j, and show that

$$\frac{1}{v_i}\frac{f'(v_i)}{f(v_i)} = \frac{1}{v_j}\frac{f'(v_j)}{f(v_j)}.$$

b. Verify that both sides of the above equation must equal the same constant, C, and that as a result

$$f(v_{i,j}) = A\,exp\left(\frac{-C}{2}v_{i,j}^2\right).$$

c. Verify that because $f(v_{i,j})$ is normalized,

$$f(0) = A = \left(\frac{C}{2\pi}\right)^{1/2}.$$

d. It can be shown that in three dimensions,

$$\frac{dn}{dv} = 4\pi n f(v) f^2(0) dv.$$

From equation 3.7 we know that the average of the square of the velocities is given by

$$\bar{v}^2 = 4\pi f^2(0)\int_0^\infty v^4 f(v)dv.$$

Solve for the constant C and verify that

$$f(v) = 4\pi v^2\left(\frac{m}{2\pi kT}\right)^{3/2}\exp\left(-\frac{mv^2}{2kT}\right).$$

2. Starting from the Maxwell-Boltzmann velocity distribution function, equation 3.13, verify that

   a. $v_{avg} = \int_0^\infty v f(v)dv = \left(\frac{8kT}{\pi m}\right)^{1/2}$

   b. $v_{rms} = \sqrt{\int_0^\infty v^2 f(v)dv} = \left(\frac{3kT}{m}\right)^{1/2}$

3. Estimate the fraction of oxygen atoms in the atmosphere at 350 km altitude that have sufficient energy to induce sputtering of Ag, of Al.

4. Estimate the amount of drag makeup fuel required for a spacecraft in a 250 km circular orbit with a 1 m^2 cross-sectional area. Assume average solar cycle conditions. How would the answer change if the operational altitude were (a) dropped from 250 km to 200 km; (b) raised from 250 km to 300 km? What if conditions were changed to (a) solar min; (b) solar max.

5. As part of the Mars Surface Exploration Program NASA intends to land a series of 200 kg instrument packages on the surface of Mars to study surface chemistry, meteorology, etc. The instrument packages will be soft-landed on the surface with parachutes. In parachute design it is more convenient to define A to be the total surface area of the parachute, $A = \pi d^2/4$, rather than its cross-sectional area when inflated. With this definition the drag coefficient is experimentally seen to be on the order of 0.55. Earth's atmospheric mass density is about 1.22 kg/m^3. Assume that the Martian atmospheric mass density is about 1% that of the Earth's and determine the diameter of parachute that will be needed if the

maximum acceptable impact velocity is 25 m/s. The acceleration of gravity at the surface of Mars is 3.72 m/s^2.

6. When its apogee kick motor fails, a satellite intended for a geosynchronous orbit is stranded in LEO in a 350 km circular orbit. Estimate the amount of time that NASA has in which to launch a rescue mission to install a new kick motor before atomic oxygen can erode the 2 mil thick silver solar array interconnects to a thickness of 1 mils. Assume average solar conditions and that the spacecraft is stabilized so that the solar arrays are always normal to the ram direction. How does the answer change if the solar arrays are always normal to the Sun?

7. An orbiting tactical communications satellite is in thermal equilibrium at 25° C. Assume the communications satellite is spherical, has a radius of 2 m, has a uniform emissivity of 0.80, and is at a distance of 5,000 km from a surveillance satellite orbiting at 250 km.

   a. Calculate the wavelength at which the communications satellite radiates most of its energy from Wien's displacement law,

   $$\lambda_{max} T = \text{constant} = 2.898 \times 10^{-3} \text{ mK}.$$

   This is the wavelength of greatest interest to the surveillance satellite.

   b. Blackbody radiation from the communication satellite is described by the Planck radiation law,

   $$E(\lambda) = \frac{8\pi hc}{e^{hc/\lambda kT} - 1} \frac{1}{\lambda^5}.$$

   Estimate the number of photons emitted by the communications satellite within a 100-angstrom band centered on the wavelength identified in part (a).

   c. Compare this with the total number of photons that could be produced near the surveillance satellite due to glow. What fraction of the total glow emission could occur within the waveband of interest before the "signal to noise" ratio of the sensor was noticeably affected?

3.8 Applicable Standards

American National Standard Guide to Reference and Standard Atmosphere Models, ANSI/AIAA G-003-1990.

3.9 References

1. Fleagle, R. G., and Businger, J. A., *An Introduction to Atmospheric Physics*, 2d ed. (Academic Press, 1980).
2. Jursa, S. A., ed., *Handbook of Geophysics and the Space Environment* (Air Force Geophysics Laboratory, Air Force Systems Command, United States Air Force, 1985).
3. Tascione, T. F., *Introduction to the Space Environment* (Malibar, FL: Orbit Book Company, 1988).
4. Mandl, F., *Statistical Physics* (New York: John Wiley & Sons, 1971).
5. Reichl, L. E., *A Modern Course in Statistical Physics* (Austin: University of Texas Press, 1980).
6. Kittel, C., and Kroemer, H., *Thermal Physics* (San Francisco: W. H. Freeman, 1980).
7. Hedin, A. E., "MSIS-86 Thermospheric Model," *J. Geophys. Res.*, 92, no. A5, pp. 4649–4662 (May 1987).
8. Walterscheid, R. L., "Solar Cycle Effects on the Upper Atmosphere: Implications for Satellite Drag," *J. Spacecraft*, 26, no. 6, pp. 439–444 (November–Dececember 1989).
9. Boring, J. W., and Humphris, R. R., "Drag Coefficients for Free Molecule Flow in the Velocity Range 7–37 km/s," *AIAA Journal*, 8, no. 9, pp. 1658–1662 (September 1970).
10. Knechtel, E. D., and Pitts, W. C., "Normal and Tangential Momentum Accommodation for Earth Satellite Conditions," *Aeronautica Acta*, 18, pp. 171–184 (1973).
11. Seidel, M., and Steinheil, E., "Measurement of Momentum Accommodation Coefficients on Surfaces Characterized by Auger Spectroscopy," in *Rarefied Gas Dynamics,* Proceedings of the 9th International Symposium 1974, vol. 2, ed. Becker and Fiebig, pp. E.9.1–E.9.12 (Germany: DFVLR Press, 1974),
12. Herrero, F. A., "The Lateral Surface Drag Coefficient of Cylindrical Spacecraft in a Rarefied Finite Temperature Atmosphere," *AIAA Journal*, 23, no. 6, pp. 862–867 (June 1985).
13. Cook, G. L., "Satellite Drag Coefficients," *Planet. Space Sci.*, 13, pp. 929–946 (1965).

14. Fredo, R. M., and Kaplan, M. H., "Procedure for Obtaining Aerodynamic Properties of Spacecraft," *J. Spacecraft*, 18, no. 4, pp. 367–373 (July–August 1981).
15. Liu, S.-M., Sharma, P. K., and Knuth, E. L., "Satellite Drag Coefficients Calculated from Measured Distributions of Reflected Helium Atoms," *AIAA Journal*, 17, no. 12, pp. 1314–1319 (December 1979).
16. Moe, M. M., and Tsang, L. C., "Drag Coefficients for Cones and Cylinder's According to Schamberg's Model," *AIAA Journal*, 11, no. 3, pp. 396–399 (March 1973).
17. Moore, P., and Sowter, A., "Application of a Satellite Aerodynamics Model Based on Normal and Tangential Momentum Accommodation Coefficients," *Planet. Space Sci.*, 39, no. 10, pp. 1405–1419 (1991).
18. Wachman, H. Y., "The Thermal Accommodation Coefficient: A Critical Survey," *ARS Journal*, pp. 336 (January 1962).
19. Bohdansky, J. J., and Roth, H.L.B., "An Analytical Formula and Important Parameters for Low-Energy Ion Sputtering," *J. Appl. Phys.*, 51, p. 2861
20. Laher, R. R., and Megill, L. R., "Ablation of Materials in the Low-Earth Orbital Environment," *Planet. Space Sci.*, 36, no. 12, pp. 1497–1508 (1988).
21. Arnold, G. S., and Peplinski, D. R., "Reaction of Atomic Oxygen with Vitreous Carbon: Laboratory and STS-5 Data Comparisons," *AIAA Journal*, 23, no. 6, pp. 976–977 (June 1985).
22. Arnold, G. S., and Peplinski, D. R., "Kinetics of Oxygen Interaction with Materials," paper 85-0472, American Institute of Aeronautics and Astronautics, 24th Aerospace Sciences Meeting, Reno, NV (1985).
23. Arnold, G. S., and Peplinski, D. R., "Reaction of Atomic Oxygen with Polyimide Films," *AIAA Journal*, 23, no. 10, pp. 1621–1626 (October 1985).
24. Leger, L. J., and Visentine, J. T., "A Consideration of Atomic Oxygen Interactions with the Space Station," *J. Spacecraft*, 23, no. 5, pp. 505–511 (September–October 1986).
25. Visentine, J. T., Leger, L. J., Kuminecz, J. F., and Spiker, I. K., "STS-8 Atomic Oxygen Effects Experiment," paper 85-0415, American Institute of Aeronautics and Astronautics, 24th Aerospace Sciences Meeting, Reno, NV (1985).
26. Slemp, W. S., Santos-Mason, B., Sykes, G. F., Jr., and Witte, W. G., Jr., "Effects of STS-8 Atomic Oxygen Exposure on Composites, Polymeric Films and Coatings," paper 85-0421, American Institute of Aeronautics and Astronautics, 24th Aerospace Sciences Meeting, Reno, NV (1985).

27. Zimcik, D. G., and Maag, C. R., "Results of Apparent Atomic Oxygen Reactions with Spacecraft Materials during Shuttle Flight STS-41G," *J. Spacecraft*, 25, no. 2, pp. 162–168 (March–April 1988).
28. Banks, B. A., Mirtich, M. J., Rutledge, S. K., Swec, D. M., and Nahra, H. K., "Ion Beam Sputter-Deposited Thin Film Coatings for Protection of Spacecraft Polymers in Low Earth Orbit," *NASA TM-87051* (January 1985).
29. Banks, B., A., Mirtich, M. J., Rutledge, S. K., and Swec, D., "Sputtered Coatings for Protection of Spacecraft Polymers," *NASA TM-83706* (April 1984).
30. Peters, P. N., Gregory, J. C., and Swann, J. T., "Effects on Optical Systems from Interactions with Oxygen Atoms in Low Earth Orbits," *Applied Optics*, 25, no. 8, pp. 1290–1298 (April 1986).
31. Gulino, D. A., "Solar Dynamic Concentrator Durability in Atomic Oxygen and Micrometeoroid Environments," *J. Spacecraft*, 25, no. 3, pp. 244-249 (May - June 1988).
32. Rudledge, S., Banks, B., DiFilippo, F., Brady, J., Dever, T., and Hotes, D. "An Evaluation of Candidate Oxidation Resistant Materials for Space Applications in LEO," *NASA TM-100122* (November 1986).
33. Yee, J. H., and Abreu, V. J., "Visible Glow Induced by Spacecraft-Environment Interactions," *Geophys. Res. Lett.*, 10, p. 126 (1983).
34. Torr, M. R., Torr, D. G., and Owens, J. K., "Optical Environment of the Spacelab 1 Mission," *J. Spacecraft*, 25, no. 2, pp. 125–131 (March–April 1988).
35. Mende, S. B., Swenson, G. R., Clifton, K. S., Gause, R., Leger, L., and Garriott, O. K., "Space Vehicle Glow Measurements on STS 41-D," *J. Spacecraft*, 23, no. 2, pp. 189–193 (March–April 1986).
36. Abreu, V. J., Skinner, W. R., Hays, P. B., and Yee, J.-H., "Optical Effects of Spacecraft-Environment Interaction: Spectrometric Observations of the DE-2 Satellite," *J. Spacecraft*, 22, no. 2, pp. 177–180 (March–April 1985).
37. Papazian, H. A., "Spacecraft Glow," *J. Spacecraft*, 24, no. 6, pp. 565–567 (November–December 1987).
38. Garrett, H. B., Chutjian, A., and Gabriel, S., "Space Vehicle Glow and its Impact on Spacecraft Systems," *J. Spacecraft*, 25, no. 5, pp. 321–340 (September–October 1988).
39. Green, B. D., and Murad, E., "The Shuttle Glow as an Indicator of Material Changes in Space," *Planet. Space Sci.*, 34, no. 2, pp. 219–224 (1986).
40. Caledonia, G. E., Krech, R. H., and Green, B. D., "A High Flux Source of Energetic Oxygen Atoms for Material Degradation Studies," *AIAA Journal*, 25, no. 1, pp. 59–63 (January 1987).

41. Cross, J. B., Spangler, L. H., Hoffbauer, M. A., and Archuleta, F. A., "High Intensity 5 eV CW Laser Sustained O-Atom Exposure Facility for Material Degradation Studies," *SAMPE Quarterly*, 18, no. 2, pp. 41–47 (January 1987).
42. Cole, R. K., Albridge, R. G., Dean, D. J., Haglund, R. F., Jr., Johnson, C. L., Pois, H., Savundararaj, P. M., Tolk, N. H., Ye, T., and Daech, A. F., "Atomic Oxygen Simulation and Analysis," *Acta Astronautica*, 15, no. 11, pp. 887–891 (1987).
43. Johnson, C. L., Albridge, R. G., Barnes, A. V., Cole, R. K., Dean, D. J., Haglund, R. F., Jr., Pois, H., Savundararaj, P. M., Tolk, N. H., Ye, J., and Daech, A. F., "The Vanderbilt University Neutral O-Beam Facility," *SAMPE Quarterly*, 18, no. 2, pp. 35–40 (January 1987).

4 The Plasma Environment

Surely the stars are images of love.
—Phillip James Bailey, Festus

4.1 Overview

In ancient Greece, the conventional wisdom agreed that all matter was composed of differing amounts of the four fundamental states of matter: earth, water, air, and fire. Over 2000 years later the conventional wisdom would agree that there are four basic states of matter, but would label them solid, liquid, gas, and plasma. The difference between the four states is a measure of the available (thermal) energy. Heating a solid produces a liquid, heating a liquid produces a gas, and heating a gas produces a plasma.[1,2] Basically, a plasma is produced when an atomic electron receives enough energy to escape the electrical attraction to the nucleus. The end result is a mixture of negatively charged electrons and positively charged atoms, which are then called ions. More formally, a plasma can be defined as a gas of electrically charged particles in which the potential energy of attraction between a typical particle and its nearest neighbor is smaller than its kinetic energy. In other words, the electrons have enough kinetic energy to remain free from the positively charged ions. Otherwise they would shortly recombine and neutrality would be reestablished. For most plasmas, the electron density n_e is equal to the ion density n_i and is simply referred to as the plasma density n_o. Over 99% of the universe, the Sun and the stars, is a plasma. Mankind just happens to live on a small piece of the one percent that is not.

Much of the orbital environment near the Earth is in the plasma state. A spacecraft that is subjected to a plasma may be charged to high electrical potentials. Because of differences in surface conductivity, conductors and dielectrics will charge to different potentials in the presence of a plasma. If the potential difference is great enough, arc discharging between the surfaces may ensue. This is of concern due to the possibility of physical damage, which could permanently damage spacecraft subsystems, or of arc-related electromagnetic interference (EMI), which could interfere with sensitive electronics. Arcing is an issue that invariably receives a great deal of attention due to the potentially disastrous consequences that may follow an arc discharge.

4.1.1 Plasma Formation

In LEO, the solar UV ionizes the ambient oxygen and nitrogen atoms, producing a plasma as illustrated in figure 4.1. Note that because the ions are essentially the same mass as the neutrals, their temperature is approximately equal to that of the neutrals (fig. 3.9). Because photoionization is the dominant production mechanism below about 600 km, this region is often referred to as the *ionosphere*. Historically, the first studies of the ionosphere that detected the ambient plasma indicated the presence of induced electric fields with the letter *E*. When the plasma was found to extend above and below the initial region of interest, they were labeled with the symbols *F* and *D*, respectively. The *F* layer of the ionosphere is where the peak in density occurs, at about 300 km altitude. Because the LEO plasma is produced when the solar UV ionizes ambient neutrals, the plasma density is seen to vary both with local time and with solar cycle as indicated below.[3]

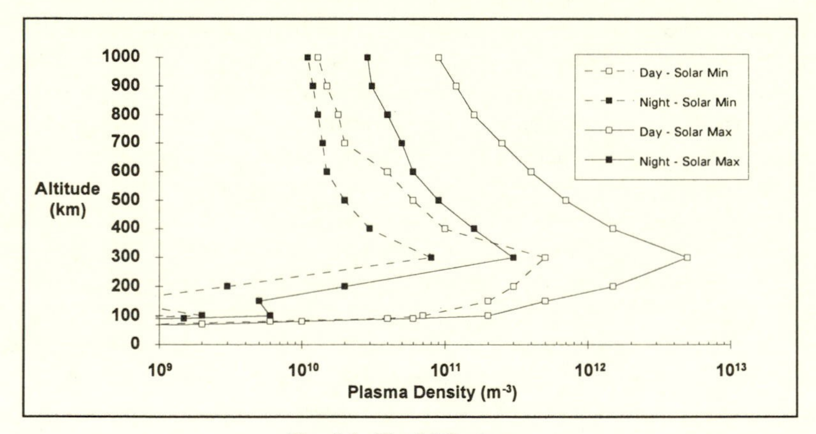

Fig. 4.1 The LEO plasma.

Above 600 km altitude the plasma is produced both by photoionization and by transport from other regions. Above 1500 km the ionized constituents are dominant over the neutrals and the region is termed the *magnetosphere*, because the local magnetic fields control the particles' motion. Plasma density at the higher altitudes is illustrated in figure 4.2. This will be reexamined in a later section.

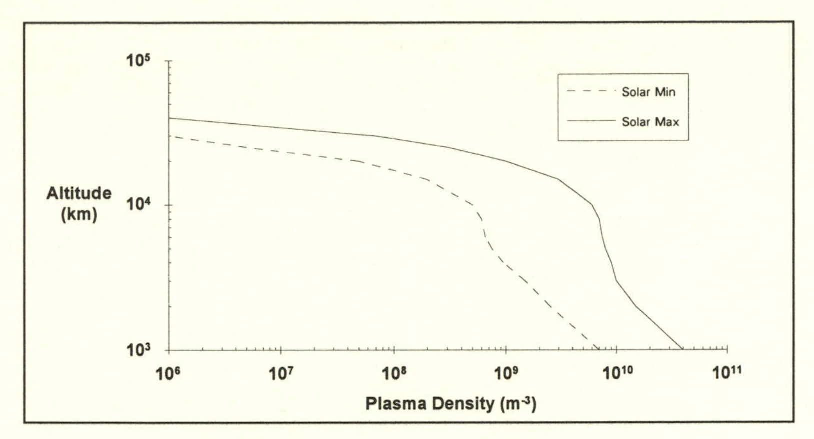

Fig. 4.2 Plasma density at the higher altitudes.

4.2 Basic Plasma Physics

4.2.1 Single Particle Motion

A charged particle moving with velocity v in the presence of either an electric or magnetic field experiences a force given by

$$\vec{F} = m\vec{a} = q(\vec{E} + \vec{v} \times \vec{B}). \tag{4.1}$$

In the absence of an electric field, if any component of v is perpendicular to B the result will be a force that acts at right angles to both v and B. Because the force acts perpendicular to the line of travel, it cannot change the magnitude of v, only its direction. Consequently, the force will be constant and will cause the particle to gyrate around the magnetic field. If B is aligned with the z-axis and v is entirely perpendicular to B, equation 4.1 reduces to

$$\begin{aligned} m\dot{v}_x &= qBv_y \\ m\dot{v}_y &= -qBv_x \\ m\dot{v}_z &= 0. \end{aligned} \tag{4.2}$$

Differentiating the first expression and substituting it into the second expression gives

$$\ddot{v}_x = -\left(\frac{qB}{m}\right)^2 v_x. \tag{4.3}$$

This expression defines the cyclotron gyration frequency, or simply the cyclotron frequency, which is given by

$$f_c = \frac{1}{2\pi}\left(\frac{qB}{m}\right). \tag{4.4}$$

The implication is that the particle must gyrate around the magnetic field lines with frequency given by equation 4.4. The constraint on circular motion is

$$a = \frac{v^2}{r}. \tag{4.5}$$

Consequently, the gyro-radius (sometimes called the Larmour radius) of a charged particle is given by

$$r = \frac{mv}{qB}. \tag{4.6}$$

In the absence of a magnetic field, an electric field can accelerate a charged particle indefinitely, in the direction of the field. If both an electric and a magnetic field are present the resultant motion is the vector sum of the forces. Consider the example of a positive test charge moving to the right in the presence of an electric field E and magnetic field B as shown in figure 4.3.

In the absence of an electric field, the $v \times B$ force would cause the particle to gyrate in a clockwise direction about the normal to the paper. However, the E field will decrease the particle's velocity as it moves downward, decreasing its gyro-radius. As the $v \times B$ force redirects the particle's velocity upward, the E field will increase its velocity and its gyro-radius. The result is that the particle is turned upward much quicker than it is turned downward. The net result is that the particle will drift to the right as shown. At any instant, the force on the particle is non-zero. However, averaged over many gyrations the acceleration, and the force, must be zero. Consequently, equation 4.1 reduces to

$$0 = q\left(\vec{E} + \vec{v}_d \times \vec{B}\right) \tag{4.7}$$

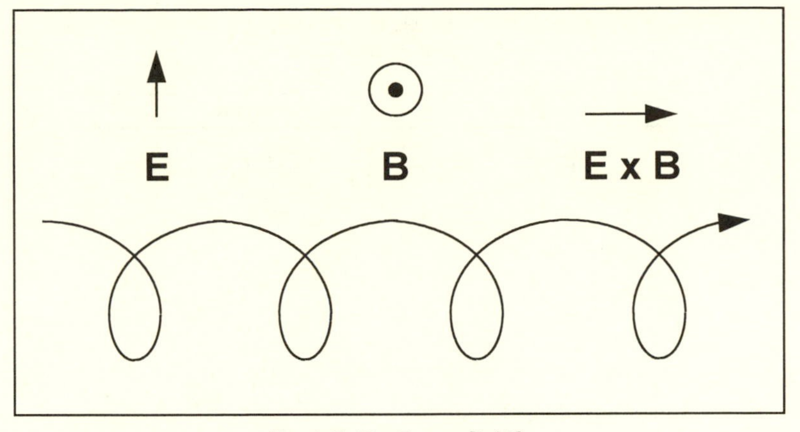

Fig. 4.3 The E cross B drift.

where v_d is the average drift velocity. It can be shown (exercise 1) that the solution for v_d is

$$v_d = \frac{\vec{E} \times \vec{B}}{B^2}. \tag{4.8}$$

This drift is known as the "E cross B" drift. Note that the preceding derivation would apply equally well to any force satisfying the constraint $\vec{F}_\perp \bullet \vec{B} = 0$. Thus, a more general expression for particle drifts is given by

$$v_d = \frac{1}{q} \frac{\vec{F}_\perp \times \vec{B}}{B^2}. \tag{4.9}$$

Table 4.1 lists the drift velocities associated with additional forces, where $v_\perp$ is the component of v perpendicular to B, $v_\parallel$ is the component of v parallel to B, and $R_{curv.}$ is the radius of curvature of the field.

We now consider the case of a charged particle moving in a magnetic field that increases in strength in the direction of the $+z$-axis as shown in figure 4.4. In cylindrical coordinates, the force on the particle at any time is given by

$$\begin{aligned} F_r &= qv_\theta B_z \\ F_\theta &= -q\left(v_r B_z - v_z B_r\right) \\ F_z &= -qv_\theta B_r, \end{aligned} \tag{4.10}$$

Table 4.1
Additional Plasma Drifts

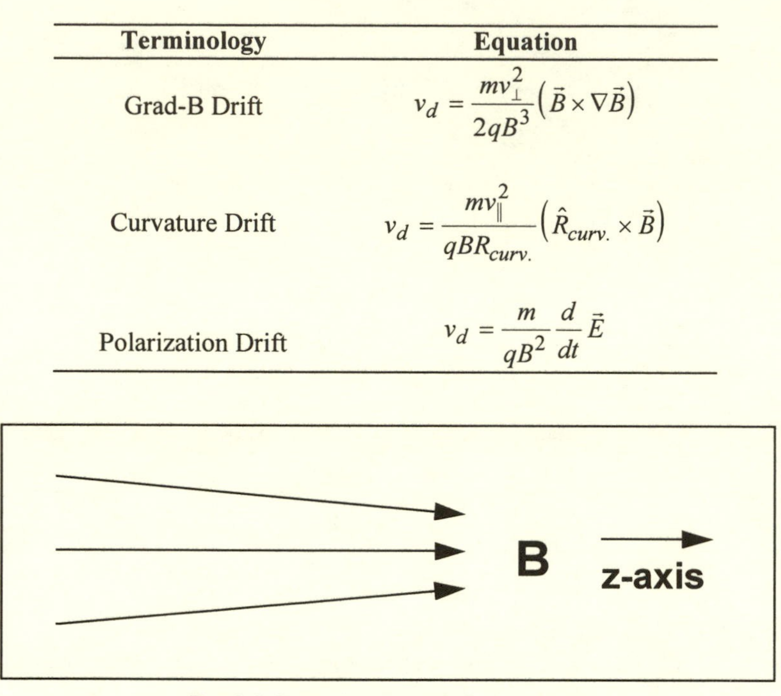

Terminology	Equation
Grad-B Drift	$v_d = \frac{mv_\perp^2}{2qB^3}\left(\vec{B} \times \nabla\vec{B}\right)$
Curvature Drift	$v_d = \frac{mv_\parallel^2}{qBR_{curv.}}\left(\hat{R}_{curv.} \times \vec{B}\right)$
Polarization Drift	$v_d = \frac{m}{qB^2}\frac{d}{dt}\vec{E}$

Fig. 4.4 Increasing magnetic field strength.

where we have assumed that $B_\theta = 0$. F_θ gives rise to the cyclotron gyration discussed previously and F_r will cause a drift in the radial direction. The expression for F_z is of greatest interest.

From Maxwell's equations $\nabla \bullet \vec{B} = 0$, which, in cylindrical coordinates, reduces to

$$\frac{1}{r}\frac{\partial}{\partial r}\left(rB_r\right) + \frac{\partial B_z}{\partial z} = 0. \tag{4.11}$$

The expression rB_r is given by

$$rB_r = -\int_0^r r\frac{\partial B_z}{\partial z}dr. \tag{4.12}$$

If $\partial B_z/\partial z$ is independent of r this constraint reduces to

$$B_r = -\frac{1}{2}r\left[\frac{\partial B_z}{\partial z}\right]. \tag{4.13}$$

Substituting this expression into equation 4.10 gives

$$F_z = \frac{qv_\theta r}{2}\left[\frac{\partial B_z}{\partial z}\right]. \tag{4.14}$$

If the force is averaged over one gyration, r is seen to be the gyro-radius, which is given by equation 4.6. Equation 4.14 reduces to

$$F_z = -\frac{mv_\perp^2}{2B}\left[\frac{\partial B_z}{\partial z}\right], \tag{4.15}$$

where $v_\theta = -v_\perp$ for a positively charged particle. The quantity

$$\mu = \frac{1}{2}\frac{mv_\perp^2}{B} \tag{4.16}$$

is defined as the magnetic moment of the particle. If the particle moves into a region of weaker or stronger magnetic field strength, its magnetic moment will remain constant even though its gyro-radius will change. Equation 4.15 can be rewritten as

$$m\frac{dv_z}{dt} = -\mu\frac{\partial B}{\partial z}. \tag{4.17}$$

Multiplying this expression by v_z on the left and the equivalent dz/dt on the right gives

$$mv_z\frac{dv_z}{dt} = \frac{d}{dt}\left(\frac{mv_z^2}{2}\right) = \frac{d}{dt}\left(\frac{mv_\parallel^2}{2}\right) = -\mu\frac{\partial B}{\partial z}\frac{dz}{dt} = -\mu\frac{dB}{dt}. \tag{4.18}$$

From equation 4.16,

$$\frac{d}{dt}\left(\frac{mv_\perp^2}{2}\right) = \frac{d}{dt}(\mu B). \tag{4.19}$$

Because energy is conserved,

$$\frac{d}{dt}\left(\frac{mv_{\parallel}^2}{2}+\frac{mv_{\perp}^2}{2}\right)=0. \tag{4.20}$$

Combining equations 4.18 and 4.19 with 4.20 gives

$$-\mu\frac{dB}{dt}+\frac{d}{dt}(\mu B)=0. \tag{4.21}$$

We must therefore conclude that $d\mu/dt = 0$, or μ is a constant.

The interpretation of this is that as the particle moves to a region of higher magnetic field, its value of $v_{\perp}$ must increase in order to keep μ constant. As a result, $v_{\parallel}$ must decrease. If B reaches a high enough value, the parallel component of v will go to zero and the particle is reflected back in the initial direction of travel. This process is called *magnetic mirroring* and is responsible for "trapping" particles in regions of weak magnetic fields. This process is used in a variety of laboratory plasma fusion experiments and is also responsible for forming the Earth's trapped radiation belts, as shown in figure 4.5.[4] Theoretically, a particle may remain trapped forever in the regions of weaker magnetic field. In practice, collisional scattering may align a particle's velocity vector with the magnetic field so that it has no magnetic moment and can escape the trapping.

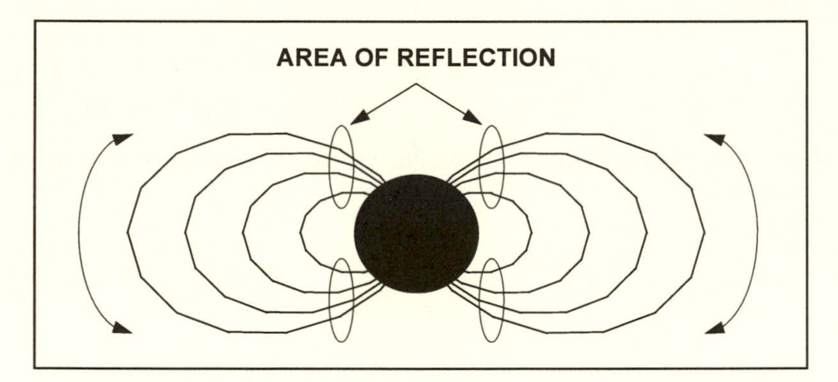

Fig. 4.5 Trapping of particles along the magnetic field lines.

4.2.2 Debye Shielding

In the absence of a plasma, the electric potential a distance r from an object having charge Q is given by

$$V = \frac{1}{4\pi\varepsilon_o}\frac{Q}{r}, \tag{4.22}$$

with ε_o being the permitivity constant. If a plasma is present near the object, the constituent of the plasma having opposite charge will be attracted to the object and will screen the potential from the rest of the plasma. Consequently, one would expect the potential to fall off with distance at some power greater than $1/r$. To see this, consider a test charge $q_t > 0$ of infinite mass that is placed in a plasma. Poisson's equation relates the electric potential $V(r)$ to the charge density. There will be three terms in the equation, corresponding to (1) the electrons, (2) the ions, and (3) the test charge.

$$\nabla \bullet \vec{E} = \nabla^2 V = -\frac{\rho}{\varepsilon_o} = \frac{e}{\varepsilon_o}(n_e - n_i) - \frac{q_t}{\varepsilon_o}\delta(\vec{r}), \tag{4.23}$$

where e (C) is the elementary charge, $n_{e,i}$ (m^{-3}) is the electron/ion density, and $\delta(r)$ is the Dirac delta function. The test charge will attract the electrons and repel the ions. This will act to increase electron density and decrease ion density near the test charge.

If we consider a gas (the plasma) in thermodynamic equilibrium at temperature T, statistical physics shows that the relation between the density of particles with energy E_2 and the density of particles with energy E_1 is

$$\frac{n(E_2)}{n(E_1)} = \exp\left(\frac{-(E_2 - E_1)}{kT}\right). \tag{4.24}$$

As the plasma approaches the test charge the electrons will gain, and the ions will lose, kinetic energy equal to $eV(r)$. Because n_o (m^{-3}) is the plasma density far from the object, this defines the average energy state and it is seen that

$$n_e(r) = n_o \exp\left(\frac{eV(r)}{kT_e}\right)$$

$$n_i(r) = n_o \exp\left(-\frac{eV(r)}{kT_i}\right), \tag{4.25}$$

where k (J/K) is Boltzmann's constant, and $T_{e,i}$ (K) is the electron/ion temperature. Performing a Taylor series expansion on equation 4.25 of the form

$$e^{-x} = 1 - x + \frac{x^2}{2!} - \frac{x^3}{3!} + \ldots, \qquad (4.26)$$

and making the assumption that $eV/kT << 1$, it is seen that away from the test charge equation 4.23 reduces to

$$\nabla^2 V = \frac{e^2 n_o}{\varepsilon_o k}\left(\frac{1}{T_e} + \frac{1}{T_i}\right)V(r). \qquad (4.27)$$

We define the Debye length for electrons and ions by the relation

$$\lambda_{e,i} = \left(\frac{\varepsilon_o k T_{e,i}}{n_o e^2}\right)^{1/2}, \qquad (4.28)$$

and the total Debye length by

$$\frac{1}{\lambda_D^2} = \frac{1}{\lambda_e^2} + \frac{1}{\lambda_i^2}. \qquad (4.29)$$

Substituting this expression into equation 4.27, Poisson's equation simplifies to

$$\frac{1}{r^2}\frac{d}{dr}\left(r^2 \frac{dV}{dr}\right) = \frac{V(r)}{\lambda_D^2}. \qquad (4.30)$$

The solution for $V(r)$ is seen to be

$$V(r) = \frac{1}{4\pi\varepsilon_o}\frac{Q}{r}\exp\left(-\frac{r}{\lambda_D}\right). \qquad (4.31)$$

As implied by equation 4.31, a few Debye lengths away from an object its potential has been effectively screened by the plasma. The Debye length can be approximated by

$$\lambda_{e,i}(m) \approx 69\left[\frac{T_{e,i}(K)}{n_o(m^{-3})}\right]^{1/2}. \qquad (4.32)$$

Example 4.1

Calculate the approximate ion Debye length associated with a 300 km orbit under average solar cycle conditions.

From figures 3.9 and 4.1, at 300 km the neutral/ion temperature is ~ 750 K and the plasma density is ~ 1×10^{12} m^{-3}.

From equation 4.31, $\lambda_i = 69(750/1 \times 10^{12})^{1/2} \sim 0.2$ cm.

4.2.3 Plasma Oscillations

One particular attribute of a plasma is what is called *collective motion.* If a few particles in a plasma are displaced, the electrical force that they exert on the other particles will cause a collective motion of the entire plasma. Consider a slab of plasma of thickness L, where the electrons are displaced a distance δ from the ions (fig. 4.6). As the result of this charge separation, an electric field will be induced that will act to pull the electrons back toward the ions (the ions are relatively immobile due to their larger masses). As a result, the electrons will accelerate back to the equilibrium position, but their momentum will cause them to overshoot until they are displaced a distance δ in the opposite direction. The plasma can oscillate in this manner at a fundamental frequency called the *plasma frequency*.

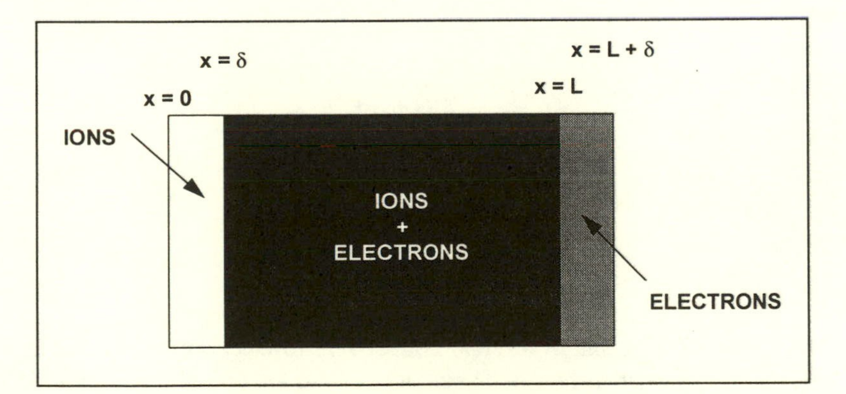

Fig. 4.6 Plasma oscillations.

In one dimension, Poisson's equation reduces to

$$\partial_x E = \frac{\rho}{\varepsilon_o}. \tag{4.33}$$

The electric field over most of the slab of plasma is dependent upon the displacement distance δ and is given by

$$E \approx -\frac{n_o e \delta}{\varepsilon_o}. \tag{4.34}$$

The force/unit area is simply the electric field multiplied by the charge per unit area, or

$$\frac{F}{A} = \left(-\frac{n_o e \delta}{\varepsilon_o} \right) (e n_o L). \tag{4.35}$$

By definition, force equals mass times acceleration so

$$\left(-\frac{n_o e \delta}{\varepsilon_o} \right) (e n_o L) = (n_o m_e L) \ddot{\delta}, \tag{4.36}$$

which reduces to

$$\left(\frac{n_o e^2}{\varepsilon_o m_e} \right) \delta + \ddot{\delta} = 0. \tag{4.37}$$

This is the equation for a simple harmonic oscillator having a fundamental frequency

$$f_{p,e} = \frac{1}{2\pi} \left(\frac{n_o e^2}{\varepsilon_o m_e} \right)^{1/2}, \tag{4.38}$$

which is called the plasma frequency. Numerically, the plasma frequency is approximated by

$$f_{p,e} \approx 9 n_o^{1/2} (m^{-3}). \tag{4.39}$$

The ability of a plasma to respond collectively to electromagnetic forces in this manner is what enables radio contact beyond the line of sight. If a radio wave of the proper frequency is aimed at a plasma, the plasma will oscillate at the same frequency and, in essence, reflect the radio wave back to the ground. This is the principle behind an ionosonde, one method of measuring electron density at orbital altitudes using ground-based radar.

Example 4.2

Calculate the approximate plasma frequency associated with a 300 km orbit under average solar cycle conditions.

From fig. 4.1, $n_o \sim 1 \times 10^{12}$ m^{-3}, so $f_{p,e} \sim 9(1 \times 10^{12})^{1/2} = 9$ MHz.

As summarized in table 4.2, in addition to plasma oscillations there are a variety of other wave phenomena that may be carried by a plasma. Although an orbiting spacecraft with an electric field detector may locate a variety of these signals at any given time, they in general have little impact on the operation of nominal spacecraft subsystems. Sensitive communications payloads are an exception, however. They may notice electromagnetic interference as a result of the natural environment.[5]

Table 4.2
Summary of Plasma Waves

Wave	Constraint
Electrostatic Waves—Electron	
Plasma Oscillations	B = 0 or v ‖ B
Upper Hybrid Waves	v ⊥ B
Electrostatic Waves—Ion	
Acoustic Waves	B = 0 or v ‖ B
Ion Cyclotron Waves	v ⊥ B
Lower Hybrid Waves	v ⊥ B
Electromagnetic Waves—Electron	
Light Waves	B = 0
O Wave	v ⊥ B and E ‖ B
X Wave	v ⊥ B and E ⊥ B
R Wave (Whistler)	v ‖ B
L Wave	v ‖ B
Electromagnetic Waves—Ion	
Alfven	v ‖ B
Magnetosonic	v ⊥ B

4.3 Spacecraft Charging

4.3.1 An Unbiased Object in LEO

The primary plasma concern in the field of space environment effects is the issue of how a spacecraft responds to the plasma environment. An object that is subjected to an unequal flux of ions and electrons will develop a net charge (fig. 4.7). Consider the example of an unbiased, conducting object placed in the LEO plasma. As shown in table 4.3, in LEO the ion thermal velocity is less than the orbital velocity, which in turn is less than the electron

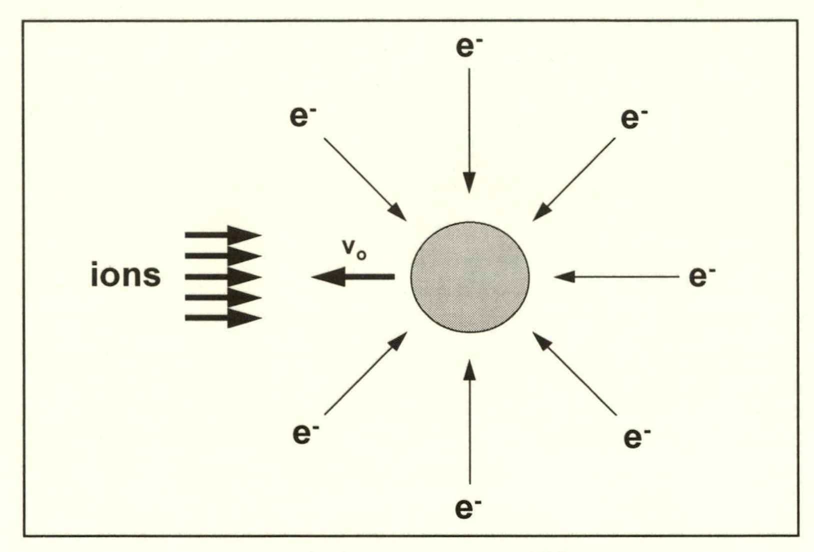

Fig. 4.7 Current collection in LEO.

Table 4.3
Ionospheric Plasma Characteristics at Shuttle Altitudes

Parameter	Symbol	Value
Altitude	h	320 km
Electron/ion density	$n_{e,i}$	1×10^5 cm^{-3}
Electron/ion temperature	$T_{e,i}$	~ 1000 K
Debye length	λ_D	< 1 cm
Electron thermal speed	$v_{e,th}$	~ 200 km/s
Ion thermal speed	$v_{i,th}$	1.1 km/s
Orbital speed	v_o	7.7 km/s
Electron plasma frequency	$f_{p,e}$	9 MHz
Electron gyroradius	$R_{L,e}$	25 cm
Ion gyroradius	$R_{L,i}$	5 m

thermal velocity. Consequently, ions may only impact those surfaces of the object facing in the direction of the velocity vector, the ram side. The ion current to the object is given by

$$I_i = en_o v_o A_i, \tag{4.40}$$

where A_i (m^2) is the area of the spacecraft that is collecting ions. Note that A_i is a function of spacecraft orientation. The electrons, on the other hand, can reach all surfaces of the vehicle, and the electron current is given by

$$I_e = \frac{1}{4} en_0 \, exp\left(\frac{eV}{kT_e}\right) v_{e,th} A_e, \tag{4.41}$$

where A_e (m^2) is the area of the spacecraft that is collecting electrons, and the factor of 1/4 arises because one-half of the electrons in the Debye sheath are actually exiting the sheath and the other half have an average component of cosθ directed toward the collecting object.[6] The spacecraft will continue to charge negatively until the potential on the spacecraft can repel the excess electrons and the currents can balance. When this occurs, the object is said to be charged to the floating potential, which is given by

$$V_{fl} = \frac{kT_e}{e} \ln\left(\frac{4 v_o A_i}{v_{e,th} A_e}\right). \tag{4.42}$$

It is easily seen that in LEO V_{fl} is typically on the order of –1 volt. Strictly speaking, the floating potential of a spacecraft refers to the potential of the conducting surfaces that are used as the spacecraft's electrical ground, measured with respect to the plasma. Note that conducting objects will charge to a global equilibrium while dielectrics will charge to a local equilibrium. Dielectric surfaces, such as solar array coverslides or thermal control surfaces, may charge to different potentials, depending on their surface conductivity and a variety of other factors that will be discussed in later sections. In LEO these potential differences will be on the order of volts.

Example 4.3

Calculate the approximate floating potential for a spherical conducting object in a 300 km orbit under average solar cycle conditions.

For a circular object, $A_i = \pi r^2$ (cross sectional area) and $A_e = 4\pi r^2$ (total surface area). Using the parameters shown in table 4.3 we find that the floating potential is given by

$$(1.38 \times 10^{-23}\ \text{J/K})(1000\ \text{K})/(1.6 \times 10^{-19}\ \text{C})\ \ln\{[(4)(8)]/[(200)(4)]\}$$

$$V_{fl} \sim -0.27\ \text{V}.$$

4.3.2 A Biased Object in LEO

We next consider the interaction of an object with surfaces biased at different electrical potentials. An important example is a solar array. As shown in figure 3.18, a solar panel is constructed by connecting individual solar cells with very thin metallic interconnects. A typical cell measures 2 × 4 cm, is covered by a protective and transparent coverslide, and is connected to its neighbors with 2-40 × 30 mil wires that are ~ 1 mil in thickness. Consequently, a typical solar panel may have ~ 0.2% of its surface area as exposed conductors. The potential difference over each cell is on the order of one volt. A number of cells will be connected in series, to form what is called a *string of cells*, in order to generate the voltage needed by the power supply. Increasing the number of strings will increase the current produced by the EPS. Each metallic interconnect in a string will be biased at a slightly different potential, relative to spacecraft ground. Consequently, different portions of the solar array will collect current from the plasma in a different manner. In order to understand the difference between this case and the case of an unbiased object, consider a solar array that is facing the ram of the plasma flow so that the metallic interconnects may all collect ion currents (fig. 4.8). This is the condition at orbital sunrise. The potential distribution along the array must arrange itself, relative to the plasma, so that the amount of ion current collected is balanced by an equal amount of electron current.

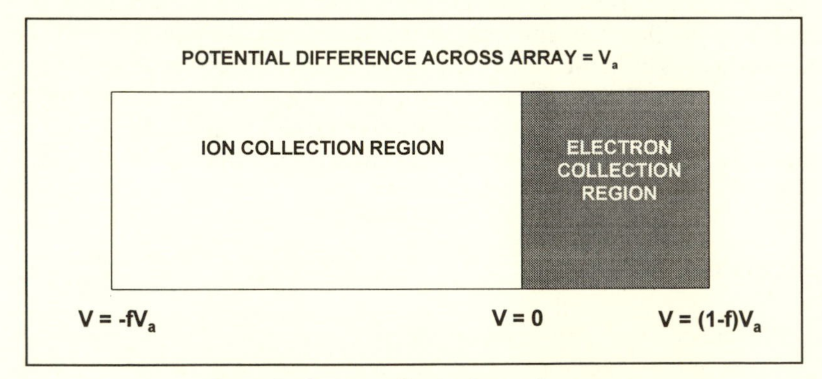

Fig. 4.8 Current collection by a solar array.

The exact solution to this problem is complicated by a variety of factors, but a fundamental understanding is easily obtained by making some simplifying assumptions. Because ions are relatively immobile, due to their larger masses, as a first approximation one can assume that ions can be collected by any portion of the array biased at a voltage less positive than the ion impact energy, ϕ_i. Portions of the array biased more positively than ϕ_i would stop and reflect the incoming ions. Similarly, electrons will be collected by any portion of the array biased less negative than their impact energy. The current density, in A/m^2, to the array is given by

$$J_i = en_o v_i \frac{fV_a - \phi_i}{V_a}$$

$$J_e = en_o v_{e,th} \frac{(1-f)V_a - \phi_e}{V_a}, \tag{4.43}$$

where f is the fraction of the array that is biased negatively, relative to the plasma, and V_a (V) is the solar array voltage. Equating J_i and J_e it is seen that the equilibrium value of f is near 1. That is, because of the relative ease with which electron current is collected, the majority of an array must float negatively in order to balance this current with the "hard-to-get" ions.

4.3.3 Spacecraft Grounding Options

Most spacecraft cannot be adequately modeled as either an unbiased object or a solar array. A spacecraft is typically a combination of the two. The floating potential of the spacecraft ground is dependent on the method used to connect the conducting surfaces of the spacecraft to the solar array. Essentially there are three options. Connecting the spacecraft to the end of the array that floats below the plasma is called *negative ground.* (Note that this configuration is sometimes referred to by the term "positive array" because the array floats positive with respect to the structures.) Connecting the spacecraft to the end of the array that floats above the plasma is called *positive ground.* Making no electrical ground at all is referred to as a *floating ground*, because both array and spacecraft float independently of each other.

If the spacecraft is grounded negatively, the spacecraft structures will contribute to ion current collection. The result is that the potential on the array will be shifted more positively with respect to the plasma. A small positive shift in array potential will be accompanied by a large increase in electron current collection by the solar array. Consequently, even though the structures may be quite large, in many cases the spacecraft will still float a significant fraction of the array voltage below the plasma. The actual value depends on the relative collecting areas of the structures and solar arrays.

If the spacecraft were grounded positively, the spacecraft structures would contribute to electron collection. As a result, the solar arrays would shift negatively to collect additional ions. The end result is that the array adds a small contribution to the effective ion-collecting area and the floating potential of the spacecraft is still quite near the plasma potential. This is due in part to the low impact energy of the electrons, 0.2 eV, as compared to the ions, 5.0 eV.

The last option, floating ground, would have no impact on the floating potential of either the structures or the array. The structures would remain biased a very few volts below the plasma potential. The differences in floating potential for the three grounding options are illustrated in figure 4.9. Note that during eclipse, when the potential difference over the solar array disappears, all three grounding configurations are equivalent.

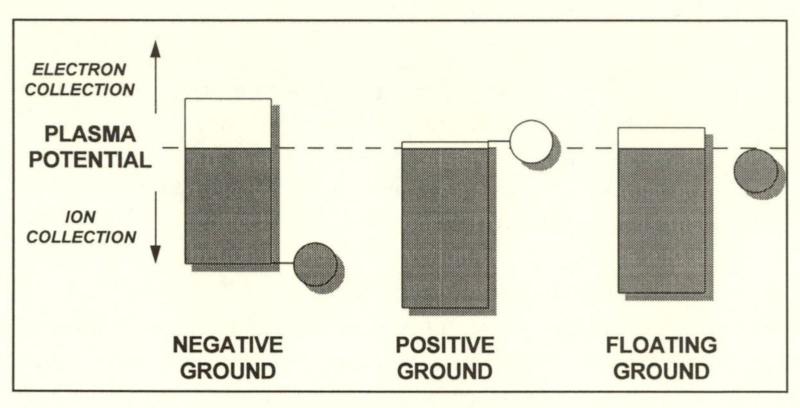

Fig. 4.9 Electrical grounding options.

From a scientific viewpoint, a positive or floating ground is preferred. A spacecraft flying plasma diagnostic instrumentation, for example, would obviously hope to pose as little disruption to the plasma as possible. Biasing the structures near the plasma potential introduces little perturbation. A negatively grounded spacecraft may pose problems for sputtering or plasma diagnostics and, as we will see, arcing. Instrumentation hoping to measure ion impact energies would find the results biased as all ions would be accelerated across the negative potential difference before reaching the instrumentation. In spite of potential drawbacks, negative ground is the usual convention because of power-system design constraints.

From an engineering viewpoint, the fundamental difference between positive and negative ground is the direction of current flow. Negative ground is the usual convention because this accommodates the flow of current through standard npn transistors. Utilizing a positive ground requires either replacing npn transistor logic with pnp counterparts, or introducing isolation techniques. For many parts required in power system design, modern technology favors npn parts over pnp parts. In order to utilize npn logic with a positive ground, the designer would have to introduce isolation techniques to accommodate the flow of current through the various electronic parts. Obviously, this complicates the design and adds to the weight of the system. Since most missions are not flying scientific instrumentation, the absolute value of the spacecraft floating potential is usually of little concern. Consequently, negative ground is the accepted convention for the majority of missions. For those cases where a positive ground would be preferred, covering the metallic solar cell interconnects would prohibit the array from collecting current and insure a floating potential near the plasma itself.

The last option, floating ground, is often avoided as the lack of a common ground makes fault detection within the electrical power system difficult. However, a floating ground does minimizes the possibility of an arc discharge damaging the power supply. Floating ground was the choice for the Russian space station *Mir* and for several U.S. interplanetary spacecraft.

4.3.4 Additional Charging Concerns

The previous discussion of spacecraft charging has ignored a variety of complicating factors that would require inclusion in a more rigorous analysis. A brief discussion of these complications is presented in the paragraphs that follow. Experience and intuition are often the best analytical tools to use in deciding whether the various phenomena warrant consideration in a spacecraft-charging analysis.

4.3.4.1 Secondary Currents

As was mentioned previously, the analysis in the preceding sections has considered only the charging currents arising from the ambient plasma itself. The thermal current flux to the spacecraft is on the order of en_0v_0. Utilizing the plasma parameters in table 4.3, one finds that this corresponds to a current density of about 1.2 mA/m^2.

Photoemission of electrons will occur when a material is illuminated with light of sufficient energy to liberate surface electrons.[7] The saturation value for the photoelectron current density of aluminum oxide is among the highest of all materials at 42 μA/m^2. Consequently, the photoelectron current is negligible, in a first approximation, for LEO orbits. However, as we will see, it may be the dominant current in the near GEO environment.[8,9]

Secondary emission of electrons can occur when ions, or neutrals, impact a material with sufficient kinetic energy to liberate electrons from the electrical attraction to the surface. Secondary emission of electrons due to proton impact on aluminum is on the order of 0.01 electrons/proton at the 5 eV impact energy found in LEO, an indication that secondary emission is also of second order in LEO.[10] At higher impact energies, which periodically occur in GEO, properly accounting for secondary emission is more critical.[11]

Finally, backscattering of incident electrons/ions can occur if materials charge to high negative/positive potentials.[12] For negative potentials, the backscatter flux is roughly 20% of the incident flux. Thus, in LEO backscattering is also negligible in a first approximation. It is of more concern in higher orbits.

A more refined spacecraft-charging analysis would include all current terms and possibly also the effects of the plasma created by hypervelocity impact of MMOD.[13] Note that the net effect of these last current terms is to make the solar array float slightly more positive. A worst case analysis for a negatively grounded spacecraft in LEO may ignore them.

4.3.4.2 Induced Potentials

From the Lorentz force law, (equation 4.1), it can be seen that an electron in a conducting rod that is moving across a magnetic field line will experience a downward force, as shown in figure 4.10. In comparison to the electrons, the ions in the rod would be essentially immobile. The electrons would respond to the $\vec{v} \times \vec{B}$ force by deflecting toward the bottom. This deflection would induce a charge separation that would give rise to an electric field. The charge separation will continue until the force due to the induced electric field equals that due to the $\vec{v} \times \vec{B}$ force and

$$\vec{E} = \vec{v} \times \vec{B}. \qquad (4.44)$$

The resulting potential difference between two points separated by a distance l is

$$V = \left(\vec{v} \times \vec{B}\right) \bullet \vec{l}. \qquad (4.45)$$

Near the Earth this equates to about 0.3 volts/m. Smaller spacecraft may usually ignore the induced potential, but a space-station-sized object would need to account for the induction in solving for global current collection.

4.3.4.3 Field-Aligned Currents

Because charged particles are constrained to follow the magnetic field lines, current collection along the magnetic field may be significantly different than current collection across magnetic field lines. For electrons in LEO, the gyro-radius is about 5 cm, smaller than most spacecraft surfaces. Consequently, most electron current collection is parallel to B. For ions the gyro-radius is about 5 m, larger than most spacecraft surfaces, and the effect is not as pronounced.

4.3.4.4 Ram/Wake Asymmetry

The filling of a wake region behind a spacecraft is similar to that of a plasma expanding into a vacuum (fig. 4.11).[14] Because of their higher mobility, the hottest electrons will enter and fill the wake region first. Because they will not be accompanied by an equal number of ions the region quickly develops a negative potential, which acts to repel the slower (cooler) electrons. Consequently, the wake region is characterized by (1) a plasma density depletion, and (2) a plasma temperature enhancement.[15,16] The wake region is defined by the Mach cone, which is bounded by the orbital velocity and the ion acoustic velocity (ion acoustic waves are to a plasma what sound waves are to a neutral gas) as shown in figure 4.12. In situ observations of the wake of the shuttle orbiter show good agreement with the Mach cone boundary (fig. 4.13).[17,18] A spacecraft that is docking, or an astronaut on a spacewalk, may become immersed in a "wake" plasma. The higher-energy plasma encountered there may give rise to more severe spacecraft charging.[19]

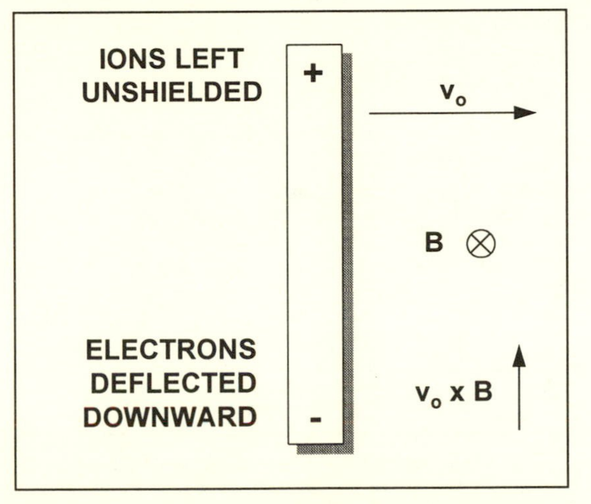

Fig. 4.10 The v cross B force displaces electrons downward.

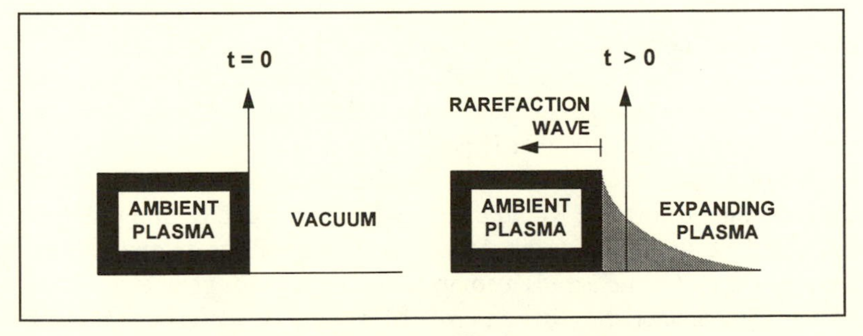

Fig. 4.11 Plasma expansion into a vacuum.

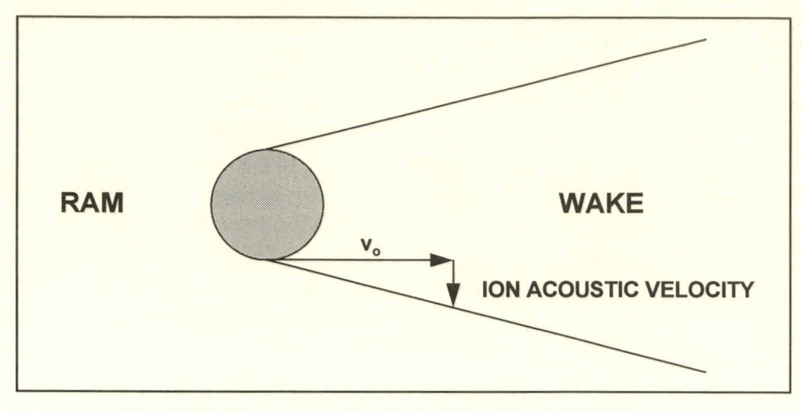

Fig. 4.12 The spacecraft wake is defined by the Mach cone.

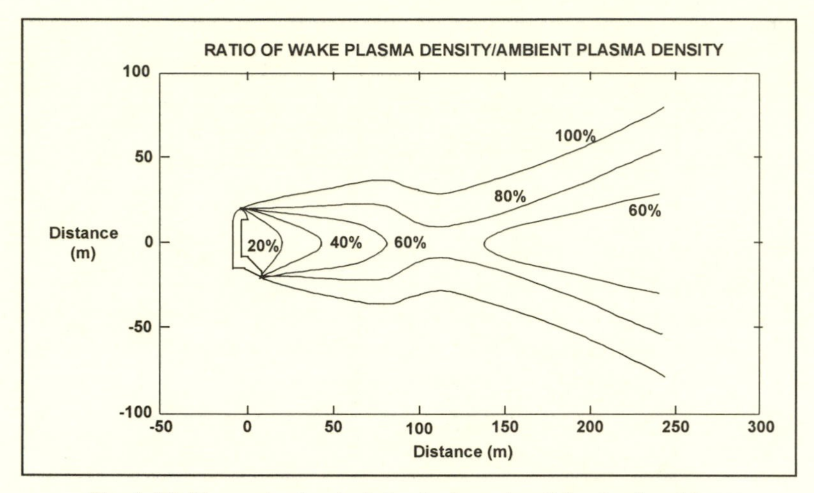

Fig. 4.13 Plasma density depletion in the wake of the shuttle orbiter.
(From Planet. Space Sci., 37, A. C. Tribble, J. S. Pickett, N. D'Angelo, and G. B., Murphy, "Plasma Density, Temperature, and Turbulence in the Wake of the Shuttle Orbiter," pp. 1001–1010, ©1989; reprinted with kind permission of Pergamon Press Ltd., Headington Hill Hall, Oxford OX3 OBW, UK)

4.3.4.5 Ion Focusing

A small localized negative potential, such as the metallic interconnects of a solar array or pinholes associated with MMOD impacts, may draw ions in from a cross-sectional area that is larger than its projected area (fig. 4.14).[20] The consequence of ion focusing is that current collection by small negatively biased surfaces would then yield a greater contribution to the total than would be thought possible based on their small cross-sectional areas. The effective collecting area is a complicated function of geometry, potential, and plasma characteristics, but an order of magnitude increase in the current collection from pinholes is possible. A small plasma-monitoring instrument, which was biased at +2000 volts, was seen to be capable of altering the floating potential of the shuttle orbiter during certain orientations.[21]

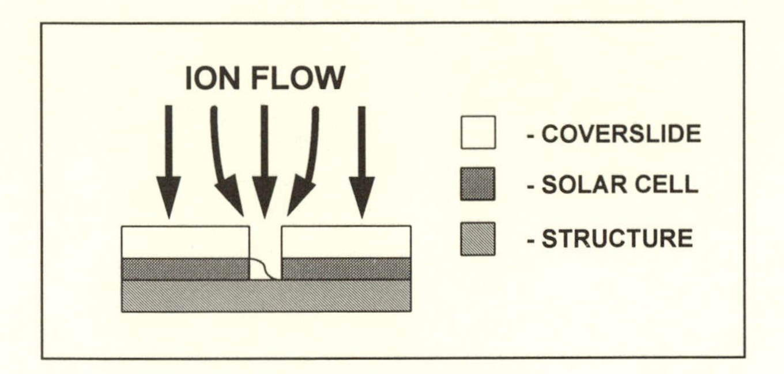

Fig. 4.14 Ion focusing.

4.3.4.6 Snapover

An additional phenomenon witnessed in the laboratory is called *snapover*. For solar arrays that are biased positively, with respect to the plasma, there appears to be a threshold voltage that, if exceeded, could cause the array to suddenly draw significant amounts of current from the surrounding plasma.[22] When the incoming electrons impact the surrounding dielectric coverslides at high energies, they are capable of inducing secondary emission of electrons, which are then in turn collected by the array. The threshold appears to be on the order of +300 volts for an array in an ionospheric plasma.

4.3.5 The Geosynchronous Environment

As shown in figure 4.2, the plasma environment encountered at the higher orbits is of much lower density than that encountered in LEO. As a consequence, a comprehensive spacecraft charging analysis in GEO must include contributions from the secondary currents discussed in the previous section. NASA specifies the parameters listed in table 4.4 for the plasma encountered in the GEO environment.[23,24] Because of the lower plasma density, under normal conditions the ambient current to a vehicle in GEO is on the order of 10 nA/m^2. Consequently, in GEO the secondary currents will dominate and spacecraft may float slightly positive. Under certain circumstances however, spacecraft in the near GEO environment can encounter energetic plasmas associated with geomagnetic storms. These storm events have been seen to charge spacecraft to as much as –20,000 volts.[25]

Table 4.4
Nominal GEO Plasma Conditions

Parameter	Units	Electrons	Ions
Number density	cm^{-3}	1	1
Current density	nA/cm^2	0.1	3.9
Energy density	eV/cm^3	3,000	11,100
Energy flux	$eV\ cm^{-2}\ s^{-1}\ sr^{-1}$	2×10^{12}	2.6×10^{11}
Population 1			
Temperature	keV	0.4	0.45
No. density	cm^{-3}	0.7	0.6
Population 2			
Temperature	keV	8.2	19.0
No. density	cm^{-3}	0.25	0.4
Avg. temp.	keV	2.4	10.0

The Sun and the Earth both have their own magnetic fields, which decrease in strength as one moves farther from their respective sources. The line marking the region where the magnetic fields are approximately the same strength is referred to as the *magnetopause*. Because the magnetic field lines from the Sun flow outward from its surface, the magnetopause flows around the Earth trailing away into space on the night side, as shown in figure 4.15. Periodically, fluctuations in the Sun's magnetic field are observed. These fluctuations result in a compression of the Earth's magnetic

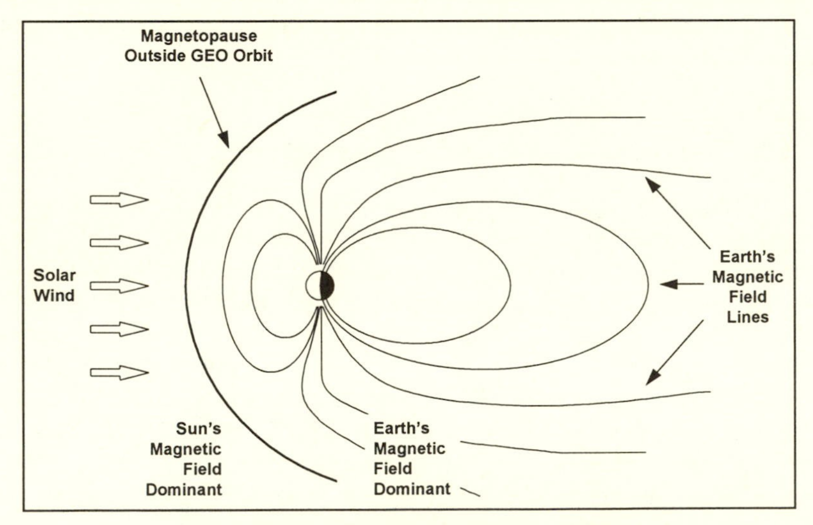

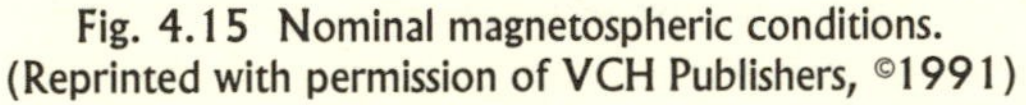
Fig. 4.15 Nominal magnetospheric conditions.
(Reprinted with permission of VCH Publishers, ©1991)

field lines, which are referred to as *geomagnetic storms*. Because the plasma is constrained to gyrate around the magnetic field lines, and to obey conservation of energy and momentum, when the compression of the Earth's magnetic field lines occurs plasma on the night side of the Earth is pushed towards the Earth's surface (fig. 4.16). As this plasma pushes closer to the Earth, the electrons and ions will be deflected by the Earth's magnetic field as shown. Because the electrons and ions are deflected in different directions, spacecraft orbiting between local midnight and 6 A.M. will see an abundance of energetic electrons. The energetic plasma encountered during this time is of much higher energy, as specified in table 4.5. Because the storm electrons are not accompanied by equal numbers of storm ions, and because it is hard for the ambient ions to keep up with the storm electrons because of their greater mass, spacecraft that are subjected to this flux can be charged to quite negative potentials. Spacecraft operating between 6 P.M. and midnight do not experience a similar effect because it is much easier for the ambient electrons to cancel the flux of storm ions. Consequently, during geomagnetic storms the greatest concern is for spacecraft operating between midnight and 6 A.M., relative to the Earth. Measurements of onboard spacecraft anomalies are shown to maximize in the midnight-to-6 A.M. corridor, in agreement with the flux of energetic plasma.[26]

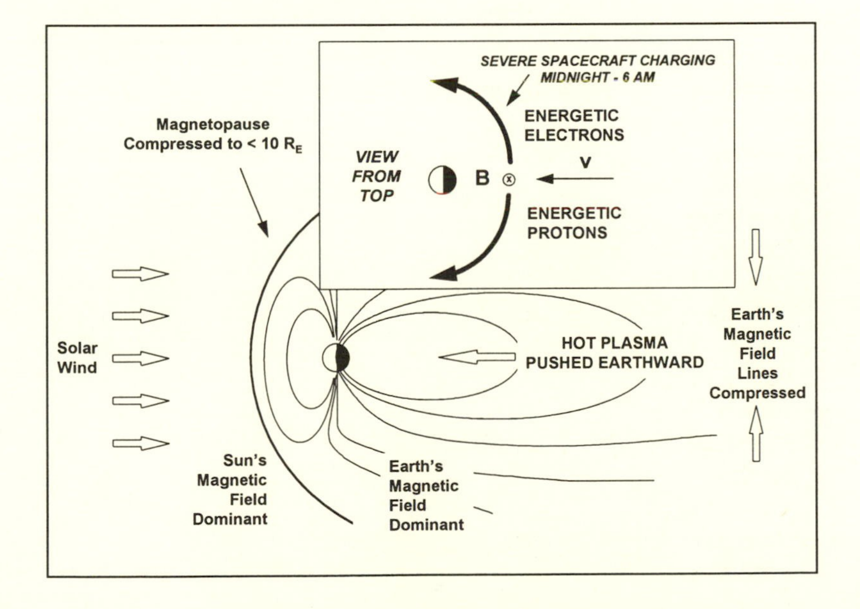

Fig. 4.16 Anomalous spacecraft charging conditions.

4.3.6 The Auroral Plasma Environment

Normally, the colder plasma found at lower altitudes is incapable of inducing significant charging. However, because energetic particles may move along the magnetic field lines, spacecraft in low-altitude polar orbits may encounter the more energetic plasma that is seen to originate at higher altitudes.[24] In situ observations confirm that auroral electrons can be accelerated to several kilovolts, producing a plasma environment capable of more severe charging (table 4.6). This energetic plasma is confined to an annular region near the poles, in the region where the magnetic field lines enter the lower altitudes. Because a spacecraft will only pass through this region periodically during the course of its orbit, charging in the auroral regions is typically of very short duration. Severe charging is more likely when the ambient plasma density is lower because the presence of the low-energy ambient plasma acts as a source of neutralizing current.

Table 4.5

Storm GEO Plasma Conditions

Parameter	Units	Electrons	Ions
Number density	cm^{-3}	1.70	1.85
Current density	nA/cm^2	0.333	0.040
Energy density	eV/cm^3	2.10×10^4	2.21×10^4
Energy flux	$eV\ cm^{-2}\ s^{-1}\ sr^{-1}$	1.61×10^{13}	1.78×10^{13}
Population 1			
Temperature	keV		
Parallel		0.50	0.27
Perpendicular		0.50	0.30
No. density	cm^{-3}		
Parallel		0.60	0.92
Perpendicular		0.50	1.00
Population 2			
Temperature	keV		
Parallel		21.7	26.7
Perpendicular		25.4	26.9
No. density	cm^{-3}		
Parallel		1.07	0.85
Perpendicular		3.40	1.45
Avg. temp.	keV	9.68	14.03

Table 4.6.
The Auroral Plasma Environment

Parameter	Value
Charging plasma energy	1–100 KeV
Charging plasma density	10^6–10^7 m^{-3}
Charging current density	100 mA m^{-2}
Background plasma density	10^8–10^9 m^{-3}
Time to traverse disturbed region	< 1 minute

4.4 Plasma Environment Effects

4.4.1 Electrostatic Discharge

In spacecraft charging analysis the concern is rarely over the absolute value of the potential itself. This is of concern only on scientific missions where instrumentation may find their readings biased by a buildup of charge on the spacecraft. These potentials may also give rise to greater drag on the vehicle, or may accelerate ions into the surface, increasing the sputtering rate.[27] In general, these effects are not mission limiting and the biggest concern is for electrostatic discharging between regions of differing potential. Formally, electrostatic discharge (ESD) can be defined as the transfer of electrostatic charge between objects at different potentials caused by direct contact or induced by electrostatic fields.

MIL-HDBK-263 defines several types of electrostatic discharge, including thermal secondary breakdown, metalization melt, bulk breakdown, dielectric breakdown, gaseous arc discharge, and surface breakdown. The first three mechanisms are energy dependent while the second three are voltage dependent. Surface breakdown is typically confined to semiconductor junction space charge regions and is mainly of concern during ground processing. Because spacecraft charging may induce large voltage differences on spacecraft surfaces dielectric breakdown and gaseous arc discharge are of greatest interest.

4.4.1.1 Dielectric Breakdown

Dielectric breakdown may occur if the potential difference across a dielectric exceeds the materials inherent breakdown characteristics.[28] The breakdown process starts with the onset of a "pre-breakdown" condition, noted by small rapid current pulses in the material, at induced electric field strengths on the order of 10^5 V/cm. Consequently, one way to avoid this type of breakdown is to maintain material thicknesses great enough to keep induced fields below ~ 10^4 V/cm. Pre-breakdown may be followed by actual breakdown where the potential difference forms a gas channel through the dielectric. The energy associated with the gas formation may cause material to be liberated from the surface, posing a possible contamination concern. Numerous dielectric breakdown events may be necessary to induce noticeable changes in the behavior of bulk surface materials such as α_s/ε. A single dielectric breakdown event at the part level may lead to permanent failure.

4.4.1.2 Gaseous Arc Discharge

Closely spaced surfaces of different electrical potential may reestablish equilibrium by forming a current path through the surrounding atmosphere. If the atmosphere is sufficiently ionized to be luminous, the discharge is sometimes referred to as *corona breakdown.* Solar arrays are of special concern for this type of arcing because of the close proximity between the dielectric coverslides and the conductive metallic interconnects. In laboratory tests solar arrays are observed to arc near the interconnects or by the corners of the dielectric coverslides where differential charging is greatest.[29] The arc rate is seen to be dependent upon plasma characteristics and solar array geometry.[30] Laboratory studies report arcing thresholds between –150 and –500 volts.[21,31,32] These arc events will be accompanied by EMI, which may interfere with sensitive electronics, or may lead to physical erosion of surfaces or actual failure of electronics.[33-35]

Although the arcing process is not well characterized, it has been suggested that arcing is the result of electric field runaway at the interface of the plasma, conductor, and solar cell dielectric. Arcing can occur when neutral molecules are desorbed from dielectric coverslides and buildup over the interconnects until the internal fields induce a flashover discharge.[36] Numerical simulations based on this mechanism support the existence of arcing thresholds at voltages on the order of –150 volts. This mechanism would also explain the observation that arcing rates are greatest immediately after arrival on orbit, when outgassing rates are greatest.[37]

4.4.1.3 Energy Dissipation

The concern from any kind of arcing is for the possibility of damage associated with the resulting energy pulse. The release of energy is not unlike the discharge of a capacitor. The amount of energy released in such a discharge is

$$E = \frac{1}{2}CV^2, \tag{4.46}$$

where C (F) is the capacitance and V (V) is the potential difference between the two surfaces. For the case of a parallel plate capacitor, C is given by

$$C = \varepsilon \frac{A}{d}, \tag{4.47}$$

where ε (F m^{-1}) is the permitivity of the medium, A (m^2) is the surface area, and d (m) is the distance of separation of the plates. Discharge energies on the order of mJ, or less, may be sufficient to damage certain components. Even if the discharge does not physically damage the arc site, the resulting EMI may propagate through the system and cause upsets in the spacecraft power supply or avionics subsystems.

4.4.2 Reattraction of Contaminants

One final area of concern is that of reattraction of contaminant ions. As discussed in section 2.2.2.2, a small percentage of the neutrals outgassed by a spacecraft may be ionized by the solar UV within a few Debye lengths of the spacecraft. If the spacecraft is charged negatively these ions may be reattracted to sensitive surfaces.[38] In higher orbits, where the Debye length is longer, this process may account for a significant portion of the total contamination to an arbitrary location. It is not expected to be a problem in LEO.

4.5 Models and Tools

A number of ionospheric models are available, such as the International Reference Ionosphere 1990 (IRI 1990). For the GEO environment the nominal/storm plasmas were specified earlier in the chapter. In performing analysis of spacecraft charging, a numerical simulation program, the NASA Charging Analysis Program (NASCAP), is available.[39] NASCAP is a particle in cell code that calculates primary and secondary currents from the plasma, trajectory of particles incident upon surfaces, electrostatic potentials around objects, and shadowing of one object by another. There are three versions of NASCAP for LEO, GEO, and auroral plasma conditions. NASCAP is the only model of its kind available and includes an extensive database of materials properties. It allows the user to quantify charging potentials and/or charging currents and is capable of modeling complex geometry. The latest NASA software tool is the Environments Work Bench (EWB). EWB was developed to support the space station program and has the capability to integrate models of the orbital environmental conditions with specific spacecraft design parameters, enabling more detailed design analysis.

4.6 Design Guidelines

As shown in table 4.7, two classes of solutions exist to minimize spacecraft charging. First, prevent the buildup of large potentials by actively balancing currents to spacecraft surfaces. Second, prevent differential charging of surfaces by insuring that the entire surface is of uniform conductivity. In order to prevent the buildup of large potentials a spacecraft may fly a plasma-generating device called a "plasma contactor" to effectively ground the spacecraft to the plasma.[40,41] By producing its own low energy plasma, a spacecraft may reattract the plasma species of opposite charge to the potential on the vehicle and minimize charging. A similar effect can be obtained from ion thrusters. Plasma contactors have been seen to limit the maximum charging potential, but they do require additional weight, power, and cost and add complexity in the design of a spacecraft.

Table 4.7
Plasma Environment Effects Design Guidelines

Uniform Surface Conductivity	Make exterior surfaces of uniform conductivity if possible
ESD Immunity	Utilize uniform spacecraft ground, electromagnetic shielding, and filtering on all electronic boxes
Active Current Balance	Consider flying a plasma contactor or a plasma thruster

Differential potentials may be minimized by making the exterior surface of a vehicle as conductive as possible. Solar arrays may be coated with transparent layers of indium oxide (IO) or indium tin oxide (ITO), which are conductive.[42,43] This was the method preferred by the International Sun Earth Explorer (ISEE) spacecraft, because the scientific instruments on the vehicle would have been affected by the buildup of differential potentials. However, this method too adds weight and expense to the vehicle design. In general, active methods such as plasma contactors or conductive coatings are only used where significant problems are foreseen.

One example where a plasma contactor was deemed necessary, as of this writing, is the space station. The space station is designing a negatively grounded EPS with 160-volt solar arrays. Without a plasma contactor the structures would, at times, float as much as 140 volts negative. The structures are utilizing a thin coating of anodized aluminum in order to maintain thermal control. Because the coatings must be thin they are unable to sustain the 140 potential difference. Consequently, arcing is expected to ensue almost immediately if a plasma contactor is not used. In order to minimize this possibility the space station is baselining a plasma contactor to maintain the floating potential less than –40 volts.

NASA recommends five general guidelines to control spacecraft charging (1) Grounding, (2) Surface Materials, (3) Shields, (4) Filtering, and (5) Procedures.[23] Grounding all conductive elements to a common ground will minimize potential differences. Similarly, making all spacecraft exterior surfaces at least partially conductive minimizes differences as well. Shielding electronics and wiring, and filtering the input to circuits, will protect the spacecraft from discharge-induced currents. Finally, proper handling, assembly, inspection, and tests to insure electrical conductivity will minimize the possibility of adverse interactions on orbit.

4.7 Exercises

1. From the definition of work,

$$W = \int \vec{F} \bullet d\vec{l} \,,$$

   verify that magnetic fields cannot change a particle's kinetic energy.

2. Make a short list of available npn transistors and their pnp counterparts. Contrast cost, power, mass, radiation tolerance, etc.

3. Walking across a carpet in very dry air can induce a potential as high as 35,000 volts. If the average human capacitance is 100 pF, calculate the energy dissipated by an electrostatic discharge that completely drains the induced voltage.

4. Consider the interaction of a spacecraft with the LEO equatorial plasma. Assume that the spacecraft (a) has a conductive structure with total surface area A_{tot} and cross sectional area A_{cs}, (b) is negatively grounded to a solar array of width w and length l, and (c) 0.2% of solar array surface area is exposed metallic interconnects. Starting from equation 4.43 and the constraint that the sum of the current to the entire system, (the structure and the solar array), must be zero, obtain a closed-form solution for the floating potential of the structure (i.e., the value of f in equation 4.43). Verify that, depending on the ratio A_{cs}/wl, the spacecraft can float negative by a significant fraction of the array voltage.

5. Consider the interaction of a space-station-like object that has as its main structural elements anodized aluminum struts with a diameter of 25 cm. Assume that there are a total of 1 km worth of struts in the entire object.

   a. Calculate the capacitance between the object and space by treating the structures as one plate of a parallel capacitor and space as the other plate. Assume the separation distance is the Debye length.

   b. If the station floats 140 volts negative, calculate the energy that could be dissipated by an arc discharge to space which shifts the potential of the object back to zero potential.

   c. How thick should the anodized aluminum coating be not to break down under an electric field strength of 10^5 V/cm? Assume a factor of safety of 2.

4.8 Applicable Standards

MIL-HDBK-263A, *Electrostatic Discharge Control Handbook for Protection of Electrical and Electronic Parts, Assemblies, and Equipment*, 20 February 1991.

MIL-STD-1541A, *Electromagnetic Compatibility Requirements for Space Systems*, 30 December 1987.

MIL-STD-1542B, *Electromagnetic Compatibility and Grounding Requirements for Space System Facilities*, 15 November 1991.

MIL-STD-1686B, *Electrostatic Discharge Control Program for Protection of Electrical and Electronic Parts, Assemblies, and Equipment*, 31 December 1992.

4.9 References

1. Nicholson, D. R., *Introduction to Plasma Theory* (New York: John Wiley & Sons, 1983).
2. Chen, F. F., *Introduction to Plasma Physics and Controlled Fusion*, 2d ed. (Plenum Press, 1984).

3. Jursa, S. A., ed., *Handbook of Geophysics and the Space Environment* (Air Force Geophysics Laboratory, Air Force Systems Command, United States Air Force, 1985).
4. Van Allen, J. A., *Origins of Magnetospheric Physics* (Washington, DC: Smithsonian Institution Press, 1983).
5. Goodman, J. M., ed., *Effect of the Ionosphere on Space Systems and Communications* (Washington DC: Naval Research Laboratory, 1975).
6. Huddlestone, R. H., and Leonard, S. L., eds., *Plasma Diagnostic Techniques* (Academic Press, 1965).
7. Lucas, A. A., "Fundamental Processes in Particle and Photon Interactions with Surfaces," in *Photon and Particle Interactions with Surfaces in Space*, ed. R.J.L. Grard, pp. 3–21 (Dordrecht, Holland: Reidel, 1973).
8. Garrett, H. B., and Forbes, J. M., "A Model of Solar Flux Attenuation During Eclipse Passage and Its Effects on Photoelectron Emission from Satellite Surfaces," *Planet. Space Sci.*, 29, no. 6, pp. 601–607 (1981).
9. Garrett, H. B., and DeForest, S. E., "Time-Varying Photoelectron Flux Effects on Spacecraft Potential at Geosynchronous Orbit," *J. Geophys. Res.*, 84, no. A5, pp. 2083–2088 (1979).
10. Whipple, E. C., Jr., "The Equilibrium Potential of a Body in the Upper Atmosphere," *NASA X-615-65-296* (1965).
11. Katz, I., Mandell, M., Jongeward, G., and Gussenhoven, M. S., "The Importance of Accurate Secondary Electron Yields in Modeling Spacecraft Charging," *J. Geophys. Res.*, 91, no. A12, pp. 13,739–13,744 (1986).
12. DeForest, S. E., "Spacecraft Charging at Synchronous Orbit," *J. Geophys. Res.*, 77, p. 651 (1972).
13. Corso, G., "Potential Effects of Cosmic Dust and Rocket Exhaust Particles on Spacecraft Charging," *Acta Astronautica*, 12, no. 4, pp. 265–267 (1985).
14. Samir, U., Wright, K. H., Jr., and Stone, N. H., "The Expansion of a Plasma into a Vacuum: Basic Phenomena and Processes and Applications to Space Plasma Physics," *Rev. Geophys. Space Phys.*, 21, no. 7, pp. 1631–1646 (1983).
15. Troy, B. E., Jr., Maier, E. J., and Samir, U., "Electron Temperatures in the Wake of an Ionospheric Satellite," *J. Geophys. Res.*, 80, no. 7, pp. 993–997 (1975).
16. Samir, U., and Wrenn, G. L., "Experimental Evidence of an Electron Temperature Enhancement in the Wake of an Ionospheric Satellite," *Planet. Space Sci.*, 20, pp. 899–904 (1972).

17. Murphy, G. B., Reasoner, D. L., Tribble, A., D'Angelo, N., Pickett, J. S., and Kurth, W. S., "The Plasma Wake of the Shuttle Orbiter," *J. Geophys. Res.*, 94, no. A6, pp. 6866–6872 (1989).
18. Tribble, A. C., Pickett, J. S., D'Angelo, N., and Murphy, G. B., "Plasma Density, Temperature, and Turbulence in the Wake of the Shuttle Orbiter," *Planet. Space Sci.*, 37, no. 8, pp. 1001–1010 (1989).
19. Katz, I., Mandell, M., Jongeward, G. A., Lilley, J. A., Jr., Hall, W. N., and Rubin, A. G., "Astronaut Charging in the Wake of a Polar Orbiting Shuttle," paper 85-7035, American Institute of Aeronautics and Astronautics, 2d Shuttle Environment and Operations Conference, Houston, TX (1985).
20. Gabriel, S. B., Garner, C. E., and Kitamura, S., "Experimental Measurements of the Plasma Sheath Around Pinhole Defects in a Simulated High Voltage Solar Array," paper 83-0311, American Institute of Aeronautics and Astronautics, 23d Aerospace Sciences Meeting, Reno, NV (1983).
21. Tribble, A. C., D'Angelo, N., Murphy, G., Pickett, J., and Steinberg, J. T., "Exposed High-Voltage Source Effect on the Potential of an Ionospheric Satellite," *J. Spacecraft*, 25, pp. 64–69 (1988).
22. Thiemann, H., and Bogus, K., "High-Voltage Solar Cell Modules in Simulated Low-Earth-Orbit Plasma, *J. Spacecraft*, 25, no. 4, p. 278 (1988).
23. Purvis, C. K., Garrett, H. B., Whittlesey, A. C., and Stevens, N. J., "Design Guidelines for Assessing and Controlling Spacecraft Charging Effects," *NASA Technical Paper 2361* (1984).
24. Davis, V. A., and Duncan, L. W., *Spacecraft Surface Charging Handbook* (Phillips Laboratory Test Report 92-2232, Air Force Materiel Command, United States Air Force, 1992).
25. Olsen, R. C., McIlwain, C. E., and Whipple, E. C., Jr., "Observations of Differential Charging Effects on ATS-6," *J. Geophys. Res.*, 86, no. A8, pp. 6809–6819 (1981).
26. Mizera, P. F., "A Summary of Spacecraft Charging Results," *J. Spacecraft*, 20, no. 5, pp. 438–443 (1983).
27. Hanson, W. B., Santini, S., and Hoffman, J. H., "Ion Sputtering from Satellite Surfaces," *J. Geophys. Res.*, 86, no. A13, pp. 11,350–11,356 (1981).
28. Frederickson, A. R., Cotts, C. B., Wall, J. A., and Bouquet, F. L., *Progress in Astronautics and Aeronautics,* vol. 107: *Spacecraft Dielectric Material Properties and Spacecraft Charging* (Washington, DC: American Institute of Aeronautics and Astronautics, 1986).
29. Snyder, D. B., "Environmentally Induced Discharges in a Solar Array," *IEEE Trans. Nuc. Sci.*, NS-29, no. 6, (1982).

30. Snyder, D. B., and Tyree, E., "The Effect of Plasma on Solar Cell Array Arc Characteristics," *NASA TM-86887* (1985).
31. Parks, D. E., Jongeward, G. A., Katz, I., and Davis, V. A., "Threshold-Determining Mechanisms for Discharges in High Voltage Solar Arrays," *J. Spacecraft*, 24, no. 4, p. 367 (1987).
32. Ferguson, D. C., "The Voltage Threshold for Arcing for Solar Cells in LEO - Flight and Ground Tests," *NASA TM-87259* (1986).
33. Leung, P., "Discharge Characteristics of a Simulated Solar Cell Array," *IEEE Trans. Nuc. Sci.*, NS-30, no. 6 (1983).
34. Adamo, R. C., and Matarrese, J. R., "Transient Pulse Monitor Data from the P78-2 (SCATHA) Spacecraft," *J. Spacecraft*, 20, no. 5, pp. 432–437 (1982).
35. Damas, M. C., and Robiscoe, R. T., "Detection of Radio-Frequency Signals Emitted by an Arc Discharge," *J. Appl. Phys.*, 64, no. 2, pp. 566–574 (1988).
36. Hastings, D. E., Cho, M., and Kuninaka, H., "Arcing Rates for High Voltage Solar Arrays: Theory, Experiment, and Prediction," *J. Spacecraft*, 29, no. 4, p. 538 (1992).
37. Bogorad, A., Bowman, C., Loman, J., Bouknight, R., Armenti, J., and Lloyd, T., "Relation Between Electrostatic Discharge Rate and Outgassing Rate," *IEEE Trans. Nuc. Sci.*, 36, no. 6, pp. 2021–2119 (1990).
38. Cauffman, D. P., "Ionization and Attraction of Neutral Molecules to a Charged Spacecraft," *Report SD-TR-80-78*, The Aerospace Corporation, Air Force Systems Command (1978).
39. Mandell, M. J., and Katz, I., "High-Voltage Plasma Interactions Calculations Using NASCAP/LEO," paper 90-0725, American Institute of Aeronautics and Astronautics, 28th Aerospace Sciences Meeting, Reno, NV (1990).
40. Hastings, D. E., "Theory of Plasma Contactors Used in the Ionosphere," *J. Spacecraft*, 24, no. 3, p. 250 (1987).
41. Lai, S. T., "An Overview of Electron and Ion Beam Effects in Charging and Discharging of Spacecraft," *IEEE Trans. Nuc. Sci.*, NS-36, no. 6, p. 2027 (1989).
42. Eagles, A. E., Amore, L. J., Belanger, V. J., and Schmidt, R. E., "Spacecraft Static Charge Control Materials," *AFML-TR-77-105* (Part 1, 1977; Part 2, 1978).
43. Eagles, A. E., and Belanger, V. J., "Conductive Coatings for Satellites," *AFML-TR-76-233* (1976).

5 The Radiation Environment

All that remained now was the distribution of the energy already released by natural laws which neither knew nor cared about the purposes of their manipulators.
—Tom Clancy, The Sum of All Fears

5.1 Overview

Shortly after the launch of *Explorer I*, a Geiger counter onboard the vehicle reported the presence of energetic charged particles, electrons and protons, trapped in orbit along the Earth's magnetic field lines. The regions of charged particles are known as the *trapped radiation belts*, or the Van Allen belts, in honor of the principal investigator who discovered them. The radiation belts differ significantly from the lower energy particles that compose the plasma environment in that the radiation belt particles are much more energetic, on the order of MeV. As a result, the radiation belt particles, and other energetic particles associated with galactic cosmic rays, solar proton events, or nuclear detonations in space, are not confined to interact with the surface of a material, but may pass through the surface layer and into the material underneath. Consequently, these particles have the capacity to cause interactions throughout a spacecraft's interior. In general, any energetic particle, (electrons, protons, neutrons, heavier ions), or photon, (gamma rays, X rays), can be considered radiation. As radiation moves through matter it may displace and/or ionize the material in its path. The affected matter itself may in turn cause further disruptions. The result is a degradation in bulk material properties. Radiation damage may decrease the power output of solar arrays, may create spurious signals in focal planes, or may induce memory errors in spacecraft avionics. Consequently, it is an area of intense interest in the field of spacecraft design.

5.2 Basic Radiation Physics

Historically, the first studies of radioactive materials defined three classes of radiation—alpha, beta, and gamma—depending on the nature of the radiation that was emitted. Today we know that an alpha particle is a helium nucleus, beta particles are either electrons or positrons, and gamma rays are energetic photons. In addition, protons, neutrons, heavy ions, and X rays can all be considered different types of radiation. As an example of the interaction between radiation and matter, an energetic electron passing through air will leave a trail of ionized particles in its wake (fig. 5.1). The air will become ionized if the energetic electron is able to strip an orbital electron free from a nucleus during its pass by an individual molecule. If the electron is moving too slowly it lacks the kinetic energy necessary to create ionizations. If it moves too fast it leaves the vicinity of a specific molecule or atom too quickly to create many ionizations. Consequently, radiation damage to a material is dependent not only on the nature of the radiation, but on the energy of the radiation and the nature of the material itself. Because radiation physics is a field that is of interest in a variety of disciplines, there are various competing terms in use to describe the amount of energy associated with radiation damage. The official SI unit of radiation is the gray, which is the amount of radiation that deposits 1 J per kg. Historically, in the field of space environment effects the terms "Rad" and "roentgen" (R) are encountered more frequently. A Rad is that amount of any kind of radiation which deposits 10^{-2} J per kg of material. A roentgen is the amount of gamma radiation (or X rays) that will produce one electrostatic unit of charge of either sign (2.08×10^9 ion pairs) in one cubic centimeter of standard temperature and pressure air. This is approximately equivalent to 0.97 Rad (table 5.1).

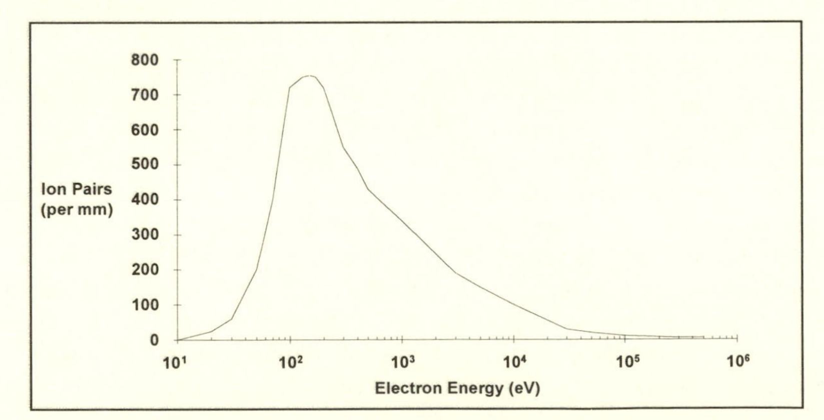

Fig. 5.1 The ionization of air by an energetic electron.

Table 5.1
Radiation Conversions

Unit	Conversion
Gray	1 J/kg
Rad	0.01 J/kg
Roentgen	0.0097 J/kg

The amount of energy deposited in a material is called the *radiation dose*. The radiation dose is dependent on the type of radiation and its energy (fig. 5.2) in addition to the material itself. For electronics, dose is often specified in Rad (Si). For biological applications, the terms Relative Biological Effectiveness (RBE) and Roentgen Equivalent in Man (REM) are often used. The RBE is the number of Rad of X ray radiation or gamma radiation that produces the same biological damage as 1 Rad of the radiation being used. The REM is defined as the product of the dose in Rad and the RBE factor. As illustrated in table 5.2, protons and fast neutrons are much more damaging than X rays or gamma rays.

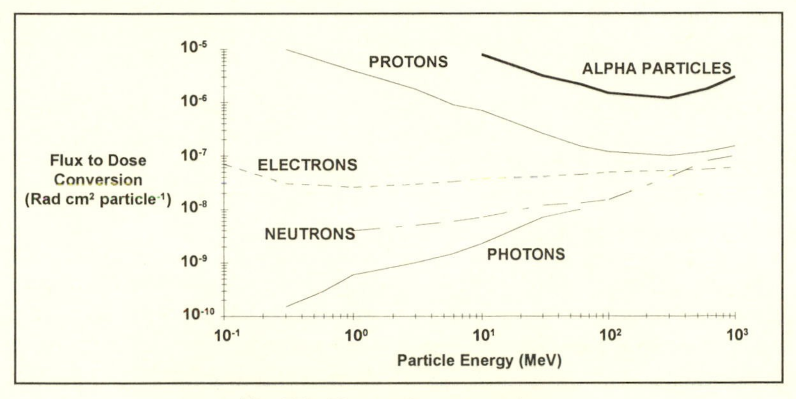

Fig. 5.2 Flux to dose conversions.

Table 5.2
Relative Biological Equivalent (RBE) Factors

Radiation	RBE
X rays, gamma rays	1
Electrons	~ 1
< 10 MeV protons	10
Alpha particles	10
Thermal neutrons	2.5
Fast neutrons	10

Many factors are important in the study of radiation damage, but most notable are (1) the total dose of radiation deposited over the life of the material, and (2) the rate at which energy is deposited, the dose rate. As shown in table 5.3, different materials have different susceptibilities to damage. Since structural materials are the least prone to damage, they are quite often used to shield more sensitive materials from damage. A visit to a radiation testing facility will usually introduce the visitor to the presence of numerous lead bricks which are used to shield the inhabitants from dangerous radiation sources. The processes that are important in the stopping of charged particles are different than those that come into play for uncharged particles. Consequently, they are best discussed separately in the following sections.

Table 5.3
Radiation Damage Thresholds

Material	Damage Threshold (Rad)
Biological matter	$10^1 - 10^2$
Electronics	$10^2 - 10^6$
Lubricants, hydraulic fluid	$10^5 - 10^7$
Ceramics, glasses	$10^6 - 10^8$
Polymeric material	$10^7 - 10^9$
Structural metals	$10^9 - 10^{11}$

5.2.1 Stopping Charged Particles

When a charged particle moves through matter it will pass close enough to the ambient atoms that electrical forces become significant. The majority of the interactions are with the atomic electrons rather than the nucleus itself because (1) the size of the nucleus, $\sim 10^{-15}$ m, is quite small in comparison to the distance at which the orbital electrons are found, $\sim 10^{-10}$ m, and (2) the electron would have to pass through the cloud of electrons to reach the nucleus. Consider the interaction between a charged particle of mass m (kg) and velocity v (m/s) and an atomic electron of mass m_e, as shown in figure 5.3. We will assume that v is so great that the approaching particle has passed before the electron has moved appreciably. At any instant, the electrostatic force between the two particles is

$$F = \frac{1}{4\pi\varepsilon_o} \frac{Ze^2}{r^2}, \tag{5.1}$$

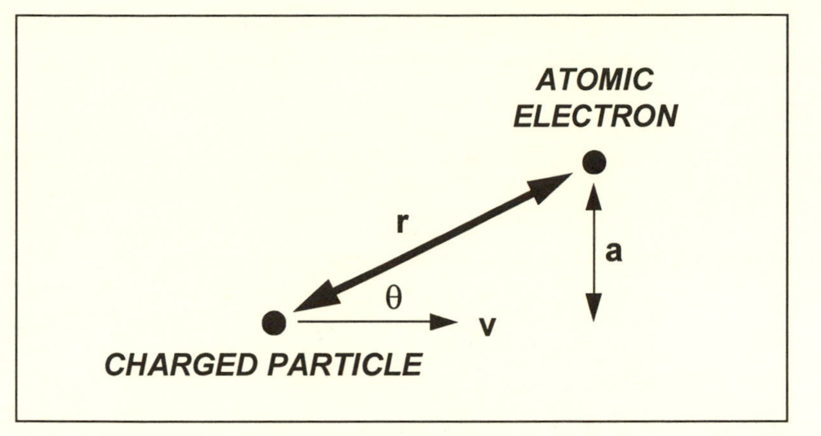

Fig. 5.3 Impact geometry.

where Ze (C) is the charge on the incoming particle and r (m) is the distance of separation. In most cases the path of the approaching particle will not directly intercept the electron; therefore the distance between the particle and the atomic electron is often expressed in terms of the distance of closest approach, a, where $r = a/\sin\theta$ with θ being the angle between the velocity vector of the approaching particle and the vector connecting the particle and the electron. The impulse given to the electron is

$$p = \int_{-\infty}^{+\infty} F\sin\theta dt. \tag{5.2}$$

If we define the coordinate axes such that the approaching particle has initial velocity aligned with the x-axis and make use of the relation $x = -a\cot\theta$, it can be shown that

$$v = \frac{dx}{dt} = \frac{a}{\sin^2\theta}\frac{d\theta}{dt}. \tag{5.3}$$

Utilizing equations 5.1 and 5.3 in equation 5.2, the momentum impulse reduces to

$$p = \int_0^{\pi} \frac{1}{4\pi\varepsilon_o}\frac{Ze^2}{av}\sin\theta d\theta = \frac{1}{2\pi\varepsilon_o}\frac{Ze^2}{av}. \tag{5.4}$$

If the velocities are small enough that relativistic effects can be ignored, the kinetic energy lost by the particle, and gained by the atomic electron, is

$$T = \frac{p^2}{2m_e} = \frac{Z^2 e^4}{8\pi^2 \varepsilon_o^2 a^2 m_e v^2}. \tag{5.5}$$

One measure of energy transfer is the stopping cross section

$$\sigma_{stop} = \int \Delta T dA, \tag{5.6}$$

where ΔT is the amount of kinetic energy lost by the particle when moving through an area dA. The cross section can be thought of as the probability of removing a particle from a beam, or as the probability of removing a given amount of energy from a single particle. In order to calculate the cross section, the limits of integration must be defined. The lower limit is defined by the Heisenberg uncertainty principle to be $h/2m_e v$, because this is the size of the electron as seen by the moving particle. The upper limit is equal to v/ν, where ν is the average of the frequencies of oscillations of the electrons in the atom, because the force exerted by the charged particle may not change the state of the electron unless the duration of the force is small in comparison with the period of oscillation.[1] Utilizing equation 5.5 in equation 5.6, the stopping cross section per electron is seen to be

$$\sigma_{stop} = \frac{Z^2 e^4}{4\pi \varepsilon_o^2 m_e v^2} \ln \frac{2m_e v^2}{h\nu}, \tag{5.7}$$

where $h\nu$ (J) is interpreted as the binding energy of electrons in the stopping medium. Obviously, this formula is not valid for lower energies because it will yield a negative cross section. Nevertheless, it is seen to qualitatively agree with experiment at higher energies. Note that the stopping cross section does not depend on the mass of the impacting particle, only its charge and velocity.

Although the concept of stopping cross section is useful, of greater interest in the study of radiation damage is the concept of stopping power. The stopping power is the amount of energy lost by a particle per unit length of path through the material. This is sometimes called *linear energy transfer* (LET) and is defined by

$$-\frac{dT}{dx} = n\sigma_{stop} = \frac{nZ^2e^4}{4\pi\varepsilon_o^2 m_e v^2}\ln\frac{2m_e v^2}{h\nu}. \tag{5.8}$$

The range of the particle in the absorbing material can be calculated by integrating equation 5.8 to give

$$R = \int dx = -\int_T^0 \frac{dT}{n\sigma_{stop}}. \tag{5.9}$$

Because equation 5.9 has an explicit dependence on the material density, it is sometimes more convenient to extract this dependence and measure the range of a particle in a material in units of length times density, or g/cm^2. Particle range in aluminum is illustrated in figure 5.4, while figure 5.5 shows the same data in units of g/cm^2. Note that the 10 MeV electron penetrates farther than the 10 MeV proton because its initial velocity is much greater.

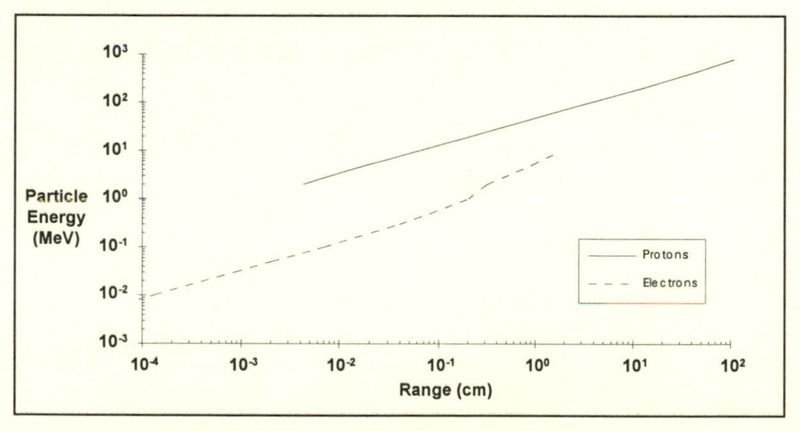

Fig. 5.4 Charged particle range in aluminum.

Because a single energetic particle may interact with numerous atomic electrons, ionizing the affected constituent atoms, the end effect may actually be a current pulse generated in the depths of the material in question. Because a key effect of this radiation is the ionizing of the ambient atoms, this radiation is sometimes called "ionizing" radiation. Less frequently, but also important in calculating radiation damage, charged particles may also displace ambient ions, disrupting the normal lattice structure.

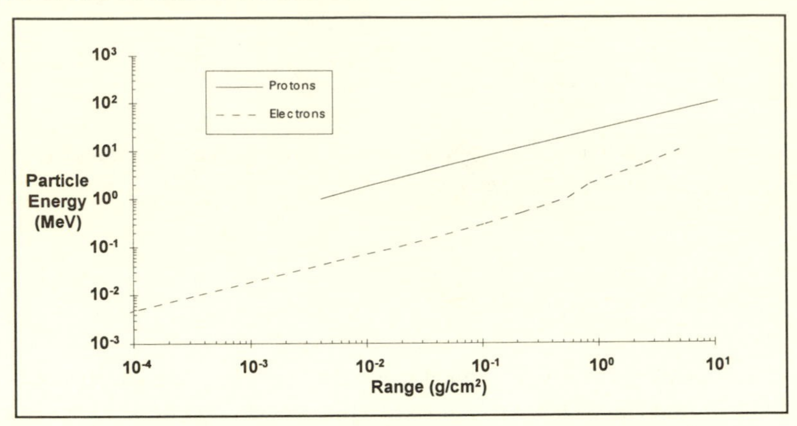

Fig. 5.5 Charged particle range in aluminum—normalized units.

5.2.1.1 Brehmsstrahlung Radiation

The preceding derivation assumed that relativistic effects could be safely ignored. At relativistic velocities charged particles may suffer an even greater energy loss when subjected to acceleration in the vicinity of the target nucleus or atomic electrons. The incoming particle may have to lose kinetic energy in the form of photons in order to satisfy constraints on conservation of energy and momentum. This radiation loss is called *brehmsstrahlung* radiation, after the German word for "braking". If two charged particles are accelerated across a potential difference, the particle with smaller mass will attain the higher velocity. Consequently, brehmsstrahlung radiation is of most concern for electrons. For relativistic electrons, those having kinetic energy much greater than $m_e c^2$, the stopping power is approximated by

$$\frac{dT}{dx} \approx -\frac{Ze^6 T(Z+1.3)[\ln(183Z^{-1/3})+0.125]}{8\pi^2 \varepsilon_o^3 h m_e^2 c^5}. \quad (5.10)$$

Because the energy loss is proportional to the kinetic energy, the electron will suffer an exponential loss of energy in the form of photons as it moves through matter. As we will see in the next section, these photons can themselves cause additional damage to the surrounding material.

5.2.2 Stopping Energetic Photons

If gamma rays or X rays are incident upon matter, they too can interact with the target material and alter the material properties. Because these rays are unaffected by electrostatic forces, they will move in straight lines until they undergo an interaction with the target material. Under these cases a beam of rays will have a probability for an interaction that is proportional to the area of the beam blocked by the target atoms divided by the total beam area. We define the total cross section σ_{tot} to be the effective target area of the nucleus. This is not to be envisioned as the actual size of the nucleus, or even a sharply defined area. Rather, if the photon and nucleus are within the radius defined by σ_{tot} they will, on average, interact. Also, while the term cross section is applied to both σ_{stop} and σ_{tot} they are different quantities. The units of σ_{stop} are energy × area, while the units of σ_{tot} are area. The relationship between σ_{tot} and the loss of photons in a material is

$$-\frac{dN}{N} = \frac{(nAdx)\sigma_{tot}}{A} = n\sigma_{tot}dx, \qquad (5.11)$$

where N is the number of incoming rays, dN is the number of interactions, n (m^{-3}) is the target number density, A (m^2) is the area of the target struck by the beam, and dx is the target thickness. The intensity of the beam will decrease exponentially as

$$N(x) = N\exp(-n\sigma_{tot}x). \qquad (5.12)$$

If one wishes to specify the absorber thickness in g/cm^2, equation 5.12 may be rewritten as

$$N(x) = N\,exp(-\mu_m\xi), \qquad (5.13)$$

where μ_m is the absorption coefficient defined by $\mu_m = n\sigma_{tot}/\rho = N_A\sigma_{tot}/M_A$, with ρ (g/cm^3) being the mass density of the target, M_A its atomic weight, N_A Avogadro's number, and ξ is the target thickness in g/cm^2. For composite materials the value of μ_m can be found from

$$\mu_m = \sum_i \frac{n_i\sigma_{tot,i}}{\rho_i}\alpha_i = N_A \sum_i \frac{\sigma_{tot,i}}{M_{A,i}}\alpha_i, \qquad (5.14)$$

where α_i is the relative abundance of the ith element.

The processes responsible for the absorption of photons (gamma rays or X rays) are (1) the photoelectric effect, (2) the Compton effect, and (3) pair production. The total cross section is the sum of the cross sections for each of these three processes.[2]

The photoelectric effect is the term used to describe the observation of a photon interacting with an atom in such a way that the photon's kinetic energy is absorbed by an atomic electron, which is then ejected. The Compton effect occurs when the gamma ray, or X ray, is scattered off of its original course of travel by an electron. Pair production occurs when an electron and a positron are created from the photon. Pair production must take place in the vicinity of a nucleus because otherwise energy and momentum cannot be conserved. Each of these processes is discussed in more detail in most nuclear physics texts. As shown in figure 5.6, the photoelectric effect is most prevalent at energies lower than 0.5 MeV, Compton scattering dominates in the range 0.5–5 MeV, while pair production is most dominant at higher energies. Because these stopping mechanisms also result in the liberation of charge carriers within the material, energetic photons such as gamma rays and X rays are also forms of "ionizing" radiation.

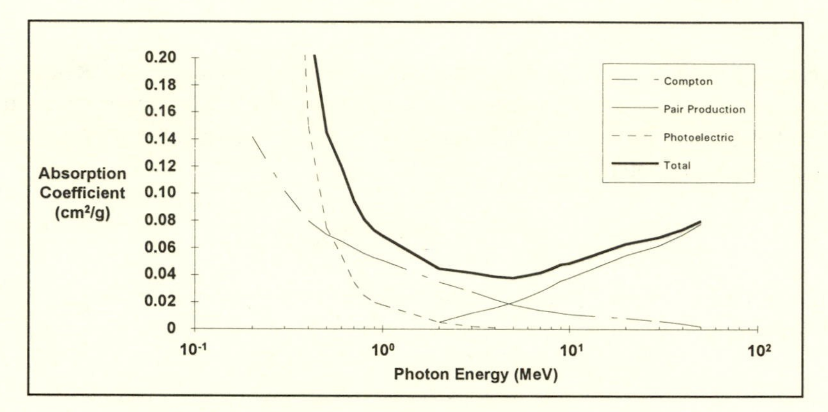

Fig. 5.6 Absorption coefficient versus photon energy.
(From Harald Enge, Introduction to Nuclear Physics (Fig. 7-6), © 1966 by Addison-Wesley Publishing Company, Inc. Reprinted by permission of the publisher)

5.2.3 Stopping Neutrons

The absorption of neutrons by matter is strongly dependent on the neutron energy. Neutrons have no electrical charge, but they do have a magnetic moment. Consequently, electromagnetic forces do come into play in stopping them. Thermal neutrons (< 1 eV) are slowed by neutron capture. Intermediate energy neutrons (eV–KeV range) have cross sections characterized by resonance structure. Fast neutrons (MeV range) are slowed by scattering with nuclei and are absorbed in neutron-capture reactions. A neutron may or may not be slowed to thermal equilibrium before capture. For fast neutrons (MeV range) the cross section is approximately three to four times the cross-sectional area of the nucleus. Eventually, the neutron will be absorbed by a nucleus, in which case the binding energy will be given off in the form of gamma rays. A good absorber of fast neutrons is paraffin. At lower energies (eV range) Cadmium has a larger cross section. As we will see in the next section, there is no significant background of naturally occurring neutron radiation in space. Consequently, this level of discussion will be sufficient.

5.3 Radiation in Space

There are three naturally occurring sources of radiation in space (1) the trapped radiation belts, (2) galactic cosmic rays (GCRs), and (3) solar proton events (SPEs). The trapped radiation belts are energetic particles, mostly electrons and protons, that are confined to gyrate around the Earth's magnetic field lines. GCRs are energetic nuclei originating outside the solar system, produced either by nova or supernova explosions in other star systems or accelerated by the interstellar fields. Finally, SPEs are energetic particles, primarily protons, that are emitted during solar flare events. Either of these sources may be dominant, depending on the spacecraft orbit.

5.3.1 Trapped Radiation Belts

As shown in chapter 4, a charged particle is constrained to gyrate around magnetic field lines. In the Earth's polar regions the magnetic field lines converge, increasing the local magnetic field strength. At the point where the magnetic field strength is

$$B = \frac{mv_{\perp}^{2}}{2\mu}, \tag{5.15}$$

a particle of mass m, velocity $v_{\perp}$, and magnetic moment μ will be reflected back in the initial direction of travel in order to conserve energy. For this reason, particles traveling along the magnetic field lines are "trapped" to gyrate back and forth along the Earth's magnetic field lines, as shown in figure 4.5. Theoretically, a particle may remain trapped forever in the magnetic field. However, by scattering, a particle may eventually be moved to higher or lower orbits or may be deflected along the magnetic field lines.

A visible confirmation of the trapping of charged particles on the magnetic field lines is the aurora. In the Northern Hemisphere the aurora borealis and in the Southern Hemisphere the aurora australialis are caused by charged particles streaming into the Earth's atmosphere along the magnetic field lines. When these charged particles interact with the neutral atmosphere they excite the ambient atoms, which in turn produce colorful radiation upon their decay. Because the magnetic field lines funnel the charged particles into the polar regions, spacecraft in polar orbits see a higher dose rate than do spacecraft in equatorial orbits. The actual size of the auroral oval varies depending on local magnetic conditions and altitude, but as a general rule of thumb, spacecraft in inclinations < 45° do not usually interact with the auroral currents.

The Earth's trapped radiation belts are comprised of two regions of electrons, peaking at about 3000 km and 25,000 km, and a single region of protons, peaking at about 3000 km, as shown in figure 5.7.[3-5] Because the radiation belts are controlled by the Earth's magnetic field, they respond to geomagnetic storms and fluctuations in solar cycle. Consequently, the actual flux of particles encountered on orbit is a function of these variables as well as magnetic latitude.

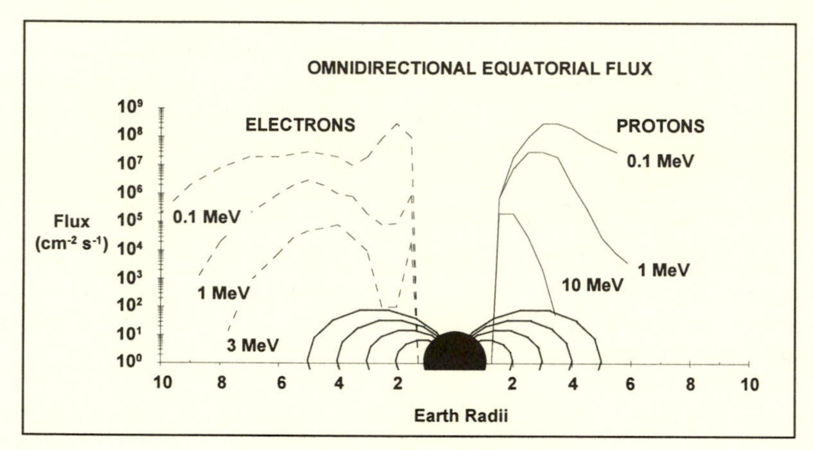

Fig. 5.7 The trapped radiation belts.

During sleep periods astronauts flying over the South Atlantic have reported seeing flashes of light in their eyes, which is sometimes described as looking like minnows splashing in a shallow bucket of water. These flashes are caused by charged particles interacting with the receptors in the astronaut's eyes. Charged particles are able to reach the lower altitudes in the South Atlantic due to an asymmetry in the Earth's magnetic field lines related to due to the displacement between the Earth's magnetic and geographical poles. This phenomena is referred to as the South Atlantic Anomaly.[3,4] Spacecraft orbiting in this region may be subjected to higher dose rates during the period of passage.

5.3.2 Galactic Cosmic Rays

Galactic cosmic radiation consists of a low flux, ~ 4 particles/cm^{2}s, of energetic, 10^8–10^{19} eV, ionized nuclei which appear to fill our galaxy isotropically. The GCR flux is composed of approximately 85% H, 14% He, and 1% heavier ions. The differential energy spectra for protons, alpha particles, and iron nuclei (for energies above 10 MeV per nucleon) is illustrated in figure 5.8.[3-5] If a GCR approaches Earth in the plane of the equator, the Earth's magnetic field will bend the particle either back to space or to the polar regions, depending on its initial direction and energy. Consequently, the Earth's magnetic field effectively shields the lower altitude/inclination orbits from some of the GCRs. Because the Earth's magnetic field responds to fluctuations in the solar output, the GCR flux is seen to be dependent on solar cycle with the GCR rate being highest at solar minimum. The dose rate due to GCRs is quite low, on the order of 3–8 Rad/year through 1–10 g/cm^2 of shielding. As we will see, the main effect of GCRs is to induce single event phenomena in electronics.

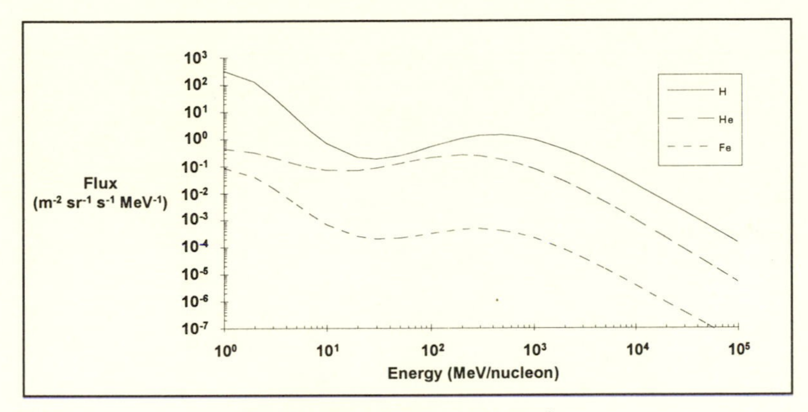

Fig. 5.8 The galactic cosmic ray flux.

5.3.3 Solar Proton Events

Periodically, the Sun is seen to eject significant amounts of protons, alpha particles, and a few heavier elements. These events, which originally were thought to correlate with solar flares, are now termed *coronal mass ejections* (CMEs). Like the GCRs, protons are the dominant constituent, and in the space radiation community these ejections are called *solar proton events* (SPEs). Most SPEs last on the order of 1–5 days, but they may subside in a few hours or continue for more than a week. As with most naturally occurring phenomena, SPEs show a range of distributions. During August 1972, and again in October 1989, there were two extremely large solar proton events. If an astronaut had been on the Moon, shielded by just a space suit, the radiation dose would probably have been lethal. To classify the size of the SPE, three "flare" models are typically used.[3-6] The ordinary model represents an average flare, the worst-case model is the 90% confidence level flare, and the anomalously large flare, based on the August

1972 flare, is the absolute worst case model. The energy spectrum of the various SPE models is illustrated in figure 5.9.

The relative effects of the natural radiation environment can be better appreciated by comparing the dose vs. depth curve for GEO (fig. 5.10) with that of a Sun-synchronous orbit (fig. 5.11). The GEO curve shows an obvious optimal design point (the knee of the curve) at ~ 1.5 g/cm^2 shielding thickness, corresponding to ~ 100 Rad/yr. Because the Earth's magnetic field would funnel charged particles into the polar region, solar protons are the major contributor to dose in Sun-synchronous orbit. (A worst-case event is assumed here.) Consequently, the Sun-synchronous orbit is more severe and shows no obvious optimal design point. This orbit would produce a more severe radiation dose on the order of 5 KRad/yr at 1.5 g/cm^2.

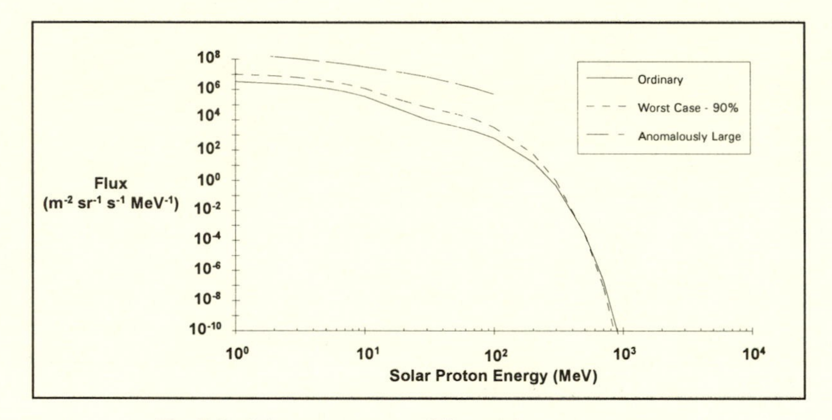

Fig. 5.9 Solar proton event differential energy spectrum.

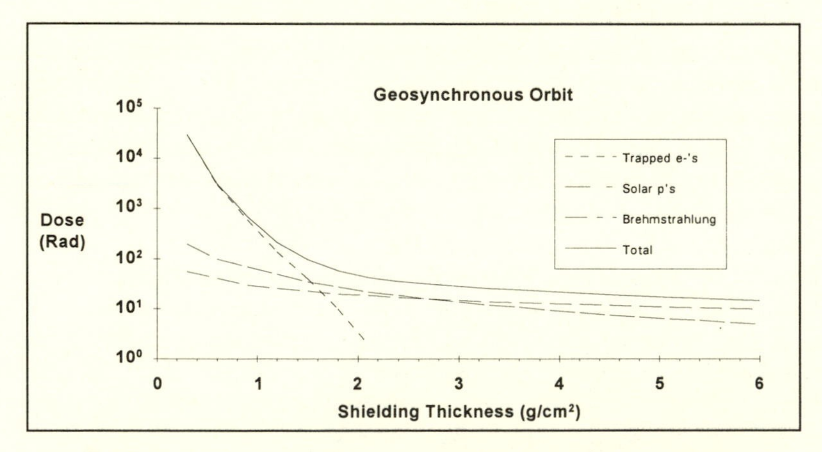

Fig. 5.10 Geosynchronous orbit dose versus depth curve for one average year.

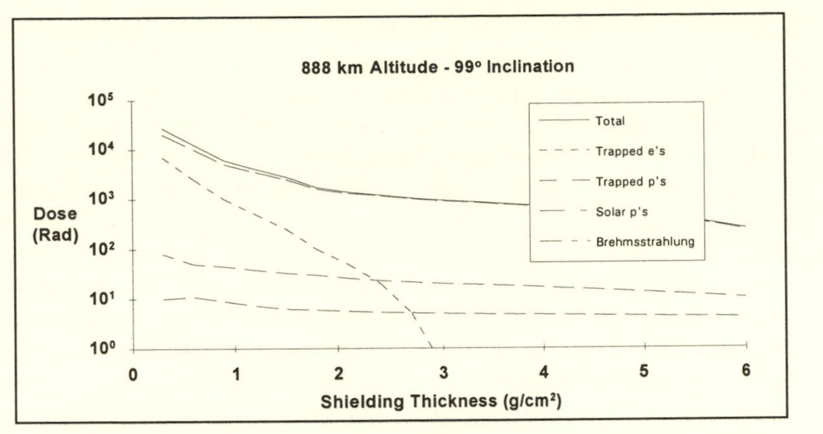

Fig. 5.11 Sun-synchronous orbit dose versus depth curve for one average year.

5.3.4 Hostile Radiation Environments

A hostile radiation environment is produced when a nuclear weapon is detonated in space. Spacecraft near the burst would be subjected to a direct flux of radiation, called the *prompt dose*, immediately following the blast. The spacecraft would be subjected to significant doses of neutron and gamma radiation, and may also be subjected to an electromagnetic pulse (EMP) that may burnout many electronic devices.[7] The immediate release of energy in the burst would strip the electrons off many ions, creating a great many charged particles. These charged particles would then become trapped by the Earth's magnetic field. Consequently, the detonation of nuclear weapons in space would also increase the number of charged particles in the trapped radiation belts. Unlike the prompt radiation, spacecraft on the other side of the Earth will be affected by the enhanced radiation in the belt. Tests of nuclear bursts in space were conducted in the 1960s. The radiation belts were significantly affected for two years after detonation, and residual effects were noticeable for at least ten years. For this reason, many military spacecraft carry additional radiation-hardening requirements to survive not only the direct effects of nuclear bursts, but also the long-term effects of the enhanced radiation belts that would result from nuclear war in space.

5.3.5 Radioisotope Thermoelectric Generators and Radioisotope Heating Units

Because the energy density that is available to a spacecraft to convert into electrical power falls off as 1/(distance from Sun squared), spacecraft that are designed for missions to the outer planets are, for the most part, unable to rely on solar arrays for power generation. Spacecraft such as

Pioneer, *Voyager*, *Galileo*, *Cassini*, *Pluto Fast Flyby*, etc., must rely on nuclear power generated onboard the spacecraft. Small nuclear units, used only for the generation of heat, are called *radioisotope heating units* (RHUs). Larger units, called *radioisotope thermoelectric generators* (RTGs), are used to generate power. These nuclear power supplies rely on the radioactive decay of nuclear isotopes (plutonium) to generate heat and electricity. Gamma rays and neutrons associated with the decay of the nuclear source can also add to the radiation budget for the spacecraft. As shown in figure 5.12, the gamma ray spectrum is fairly broadband while the neutron spectrum peaks in the 1–10 MeV range.[8] Because RTGs/RHUs are specifically tailored for each new mission that arises, it is difficult to draw generalities concerning their effect on mission requirements. However, because interplanetary spacecraft have a significant concern for the effects of solar proton events and galactic cosmic rays, RTGs and RHUs are usually not the driving source in the radiation budget.

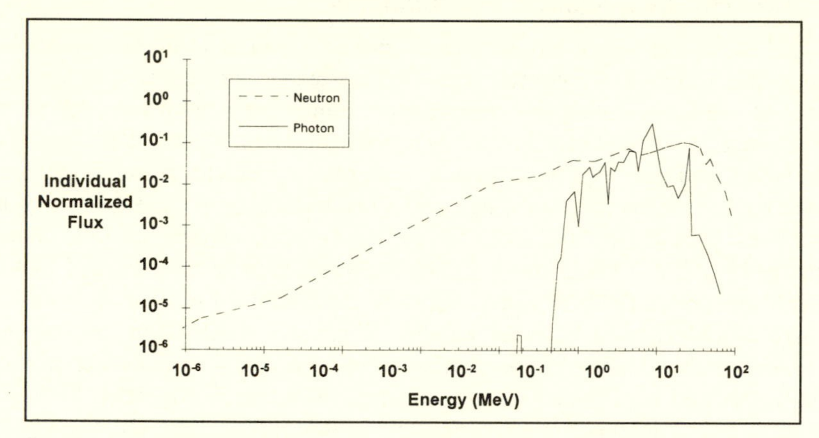

Fig. 5.12 Radiation spectrum from a typical radioisotope thermoelectric generator.

5.4 Radiation Environment Effects

5.4.1 Total Dose Effects

5.4.1.1 Solar Cell Degradation

If a photon of the proper frequency is allowed to enter an np semiconductor, the photon may be absorbed by an atomic electron, which will then have sufficient kinetic energy to escape the electrical attraction to its nucleus. For silicon, the intrinsic ionization potential is 3.6 eV, which corresponds to a wavelength of 0.345 μm (equation 2.5). Consequently, a solar cell is simply an np (or pn) semiconductor called a photodiode.

Exposing a solar cell to the Sun's visible radiation would produce a number of ionizations per unit volume. Blue light is 99% absorbed within 0.2 microns of the surface, in the n-type material thickness. Red light must travel about 200 microns, into the p-type material thickness, before being 99% absorbed (fig. 2.10). Once the photon energy has been used to ionize atoms in the material, the resulting electrons will act as free charge carriers and can contribute to current flow. The majority of the electrons are liberated deep within a cell and diffuse through the material until they are either recaptured by an ionized atom or reach the np junction. If the electron is recaptured, the energy used to generate it is lost and appears as heat. If the electron reaches the junction, however, it will be accelerated into the n-type material. Because the n-type material has relatively few atoms capable of capturing the electron, the electron will in all probability continue to move toward the surface of the cell. If electrical leads are placed on the surface, this flow of electrons can then be drawn off as current flow, which can in turn be used to power a spacecraft.[9,10] As previously mentioned, a single cell generates a potential difference of about 1 volt and a current of a few mA. By connecting the cells in series, a 28-volt potential difference can be generated, and by increasing the number of "strings" of cells additional current is produced. Typical efficiencies for silicon solar cells are about 11.5%. That is, 88.5% of the incident solar energy is lost in the form of heat.

As radiation interacts with a solar cell it will produce ionizations and atomic displacements. If an orbital electron in the coverslide becomes ionized, it may diffuse through the coverslide and eventually become trapped by an impurity atom to form a charged defect called a *color center*.[11,12] As a result, transparent polymers become darkened. If an ambient silicon atom is displaced, the structure of the lattice is altered and the average distance that an electron can diffuse before being deflected by the lattice defect is decreased. As the diffusion length, the mean free path, decreases, fewer electrons can make it from the interior of the cell to the np junction. Consequently, the current and power that a cell can produce are decreased. As shown in figure 5.13, radiation reduces not only the current, but also the voltage drop, produced by a given cell.

In order to determine the effect of radiation on a particular solar cell, it is necessary to know both the differential energy spectrum of the radiation and the response characteristics of the cell. For ease of comparison, the differential fluence is typically converted into equivalent fluences of either 1 MeV electrons, or 10 MeV protons. That is, the damage produced by a distributed spectrum of particles is equated to a number of monoenergetic particles that would be required to produce the same damage. The relative damage coefficients for electrons/protons of different energies is shown in figure 5.14. Experimentally, we can see that one 10 MeV proton produces the same damage as about three thousand 1 MeV electrons. This relationship allows one to then compare proton damage to electron damage. Note that the results shown in figures 5.13 and 5.14 are dependent on the specific design characteristics of the solar array in question. Although the discussion to this point has focused on silicon technology, solar cells can also be formed from GaAs and InP. While these technologies offer superior radiation resistance and increased efficiency, additional manufacturing costs require the designer to perform a system level trade to determine which technology is most appropriate for the orbital environment in question.[13,14]

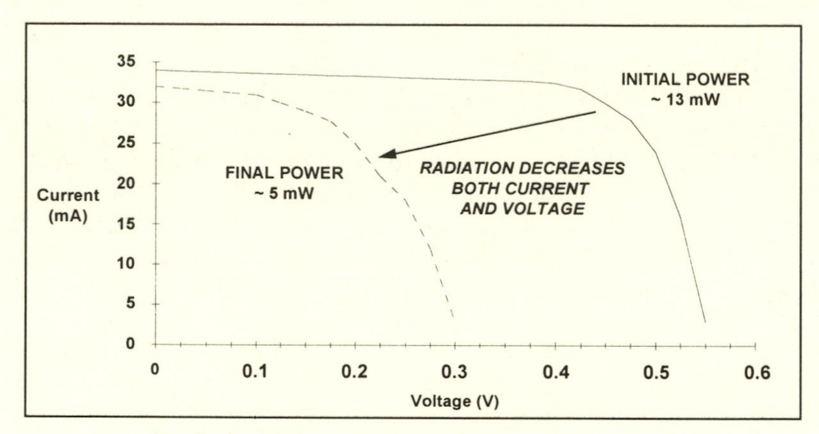

Fig. 5.13 Effect of radiation dose on power production.

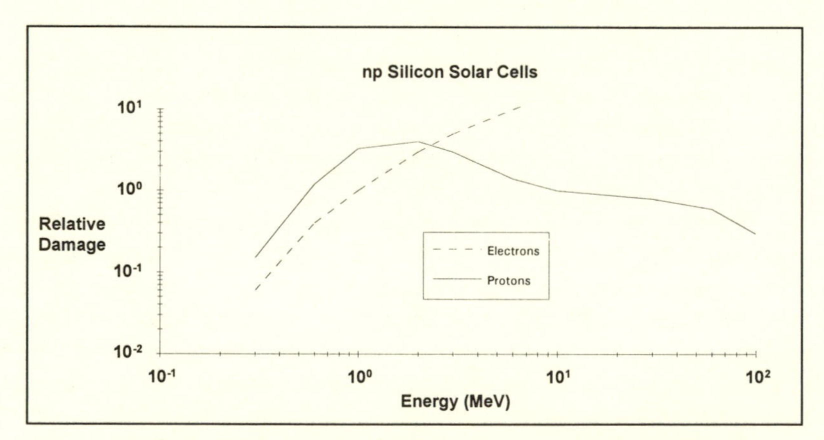

Fig. 5.14 Relative damage coefficients for protons and electrons.

5.4.1.2 Degradation of Electronics

Much like solar cells, other electronic devices rely on the diffusion of charge carriers through semiconducting material in order to operate properly. Diffusion lengths for semiconductors (~ 1 μm) may be an order of magnitude smaller than those associated with solar cells, making them especially susceptible to radiation damage. As the diffusion length of charge carriers in these devices is decreased, they too will eventually cease to function. Table 5.4 illustrates the total dose threshold for a variety of technologies. Depending on the radiation environment anticipated for a given mission, the choice of piece parts that are capable of surviving the mission may be very small. In addition to electronics, radiation can also degrade the mechanical/electrical/thermal properties of various aerospace polymer films.[15-18] The specific effect varies depending on the type of radiation, dose rate, and total exposure, with some materials incurring damage on the order of 10^4 Rad, while other materials remain usable to 10^8 Rad (table 5.5).

Table 5.4
Total Dose Thresholds for Various Electronic Technologies

Technology	Total Dose Rads (Si)
CMOS	$10^3 - 10^6$
MNOS	$10^3 - 10^6$
NMOS	$10^2 - 10^4$
PMNOS	$10^3 - 10^5$
ECL	10^7
I^2L	$10^5 - 10^6$
TTL/STTL	$> 10^6$

Table 5.5
Total Dose Effects on Aerospace Polymers

Material	Usable Dose (Rad)	Limited Use Dose (Rad)
Teflon	2×10^4	3×10^4
Nylon	3×10^5	5×10^5
Polyethylene	1×10^7	6×10^7
PVC	1×10^7	1×10^8
Mylar	3×10^6	7×10^7
Kapton	2×10^8	2×10^9
Polystyrene	7×10^8	3×10^9

5.4.1.3 Effects on Humans

Like electronics, living matter is also degraded by radiation. The actual physics of the degradation process is much more complicated, but basically the ability of a cell to reproduce properly is changed as the incident radiation interacts with a cells DNA and RNA. These changes imply that latter generations of cells may evolve into something significantly different from a "normal" cell and the organism ceases to function as efficiently. As changes propagate from generation to generation, measured in cell reproductive lifetimes, the organism may eventually lose its ability to reproduce and die. Cells which reproduce rapidly, such as blood marrow, or younger beings which are still growing, are more adversely affected.[19-21]

Table 5.6 illustrates the National Academy of Sciences' recommended dose limits for astronauts. Measured radiation dose values inside space vehicle crew compartments (tables 5.7 and 5.8)[22-24] run little danger of exceeding these values even though they are a significant enhancement over nominal ground sources (table 5.9). Conversely, surfaces on the Long Duration Exposure Facility (LDEF), which spent 5 years 9 months in LEO, saw a total dose of about 500 KRads.[25] Interplanetary missions, with lifetimes measured in years, may see even larger radiation doses, as the Earth's magnetic field would not be present to shield the spacecraft from the GCRs and SPEs. Consequently, additional shielding would be needed to insure that the crew is not susceptible to the effects of high radiation doses (table 5.10). Interplanetary missions would require such a large amount of consumables that a properly designed spacecraft should be able to rely on their shielding ability in addition to that of the spacecraft structures itself to maintain crew safety.[26]

Table 5.6
Recommended Radiation Dose Limits for Astronauts

	Dose Limits – Rads (Tissue)		
Mission Duration	**Skin (0.1 mm)**	**Eyes (3 mm)**	**Bone Marrow (5 cm)**
30 days	75	37	25
90 days	105	52	35
180 days	210	104	70
1 year	225	112	75
Career total	*1200*	*600*	*400*

Table 5.7
Radiation Dose Values for the Gemini, Apollo, and Skylab Programs

Dose (Rad)								
Gemini			Apollo			Skylab		
Mission	High	Low	Mission	High	Low	Mission	High	Low
3	0.031	0.020	7	-	-	1	-	-
4	0.050	0.042	8	0.17	0.15	2	1.81	1.62
5	0.185	0.167	9	-	-	3	4.21	3.67
6	0.025	0.024	10	0.66	0.43	4	8.02	6.80
7	0.166	0.161	11	0.20	0.18			
8	0.010	0.010	12	1.20	0.73			
9	0.022	0.017	13	-	-			
10	0.768	0.685	14	0.98	0.91			
11	0.030	0.027	15	0.32	0.28			
12	0.020	0.020	16	0.49	0.36			
			17	0.51	0.37			

Table 5.8
Radiation Dose Values for the Shuttle Program in the 1980s

Mission	Launch Date	Altitude (km)	Incl. (deg)	Duration (hrs)	Dose (Rad) High	Dose (Rad) Low
1	12 Apr 81	269	40.3	54.3	-	-
2	12 Nov 81	254	38.0	54.2	0.018	0.006
3	22 Mar 82	280	38.0	192.2	0.047	0.042
4	27 Jun 82	296	28.5	169.2	0.041	0.038
5	11 Nov 82	283	28.5	122.2	0.025	0.022
6	04 Apr 83	293	28.5	120.4	0.027	0.024
7	18 Jun 83	296	28.5	146.4	0.046	0.043
8	30 Aug 83	287	28.5	145.1	0.041	0.038
41-A	28 Nov 83	250	57.0	247.8	0.141	0.119
41-B	03 Feb 84	296	28.5	191.3	0.058	0.052
41-C	06 Apr 84	498	28.5	167.7	0.689	0.489
41-D	30 Aug 84	315	28.5	144.9	0.053	0.051
41-G	05 Oct 84	259	57.0	197.4	0.092	0.084
51-A	08 Nov 84	352	28.5	191.7	0.159	0.088
51-C	24 Jan 85	333	28.5	73.6	0.041	0.035
51-D	12 Apr 85	454	28.5	168.0	0.472	0.303
51-B	29 Apr 85	352	57.0	168.1	0.160	0.127
51-G	17 Jun 85	370	28.5	169.7	0.152	0.105
51-F	29 Jul 85	315	49.5	190.8	0.167	0.112
51-I	27 Aug 85	343	28.5	170.3	0.120	0.099
51-J	03 Oct 85	509	28.5	97.8	0.513	0.329
61-A	30 Oct 85	324	57.0	168.7	0.139	0.112
61-B	26 Nov 85	380	28.5	165.1	0.171	0.125
61-C	12 Jan 86	324	28.5	146.1	0.075	0.065
26	29 Sep 88	311	28.5	97.0	0.037	0.036
27	02 Dec 88	459	57.0	105.1	0.173	0.137
29	13 Mar 89	317	28.5	119.7	0.048	0.042
30	04 May 89	311	28.9	97.0	0.029	0.028
28	08 Aug 89	306	57.0	122.0	0.065	0.057
34	18 Oct 89	306	34.3	119.7	0.043	0.038
33	22 Nov 89	537	28.5	120.1	0.601	0.484

Table 5.9
Typical Examples of Radiation Exposure

Source	Dose Rate (Rad/yr)
Internal Irradiation	
Potassium 40	0.020
Carbon 14	0.001
External Irradiation	
Soil	0.043
Radon in air	0.001
Cosmic Rays	
Seal Level	0.035
1.5 km	0.04 – 0.06
3.0 km	0.08 – 0.12
4.5 km	0.16 – 0.24
6.0 km	0.30 – 0.45
Older Technologies	
Radium watch dial	0.04
Shoe X ray fitting	< 0.001

Table 5.10
Acute Dose Effects on Humans

Dose (Rads)	Probable Effect
0–50	No obvious effects, blood changes
80–120	10% chance of vomiting/nausea for 1 day
130–170	25% chance of nausea, other symptoms
180–220	50% chance of nausea, other symptoms
270–330	20% deaths in 2–6 weeks, or 3 mo. recovery
400–500	50% deaths in 1 mo., or 6 mo. recovery
550–750	Nausea within 4 hours, few survivors
1000	Nausea in 1–2 hours, no survivors
5000	Immediate incapacitation, death within 1 week

5.4.2 Dose Rate Effects

In addition to the atomic displacements, which decrease the diffusion length of electrons in electronic devices, radiation will also produce a certain amount of ionizations. The number of ionizations produced is dependent upon the ionization potential of the material and its density. The dose of radiation received, D (Rad), times the density of the material, ρ (g/cm^3), divided by the ionization potential, IP (J), is the measure of the number of free electrons produced per unit volume. A convenient term used in the discussion of radiation effects is the carrier generation constant g, defined by

$$g = \frac{\rho}{IP}. \tag{5.16}$$

The carrier generation constant has units of electrons per cm^3 per Rad, or equivalently, hole-electron pairs per cm^3 per Rad. For silicon the value of g is 4.05×10^{13} electrons/cm^3 Rad, while for germanium and gallium arsenide the numerical value is 1.2×10^{14} and 6.9×10^{13}, respectively. It follows from the definition of g that the ionization current produced in the region of an np junction by the passage of radiation is dependent not on the dose D, but the dose rate, $\gamma = dD/dt$, according to the relation

$$I = eAWg\gamma, \tag{5.17}$$

where e (C) is the charge on the electron, A (m^2) is the area of the junction, and W (m) is the junction width. If the radiation dose rate is high enough, the current produced by the radiation may exceed the nominal currents in the device. Consequently, the nominal signal may be swamped and the device may cease to function properly.[9,27]

5.4.2.1 Single Event Effects

Many modern electronic devices have dimensions so small that the dose rate currents arising from the passage of a single energetic particle may be sufficient to alter the operating characteristics of the device in question. The resulting disruptions are called *single event phenomena* (SEP) or *single event effects* (SEE). An effect is classified as "soft" if the damage is transitory and the device can recover. An example of a soft error is the reversible flipping of a memory bit. An effect is classified as "hard" if the damage is permanent and the device is lost, such as an irreversible bit flip. Two specific examples of SEE warranting further discussion are *latchup* and *upset*.

Latchup is said to occur when the device is transformed to an anomalous state that no longer responds to input signals. Latchup is usually confined to bulk complementary metal oxide semiconductor (CMOS) devices. For typical integrated circuits the latchup threshold is on the order of 10^8 Rad/s. Upset occurs when a device is caused to function in a manner that is not consistent with its design characteristics. For example, the resulting localized electric fields and currents associated with radiation-induced currents may cause a memory register to change its state. The register is said to have been upset by the radiation. Upset thresholds are dependent on the design specifics of the device in question and may be as low as 10^7 Rad/s or as high as 10^{12} Rad/s. Upsets resulting from the passage of a single particle of radiation are termed *single event upsets* (SEUs). One characteristic of SEUs is that they are statistically guaranteed to occur in any device that proves to be susceptible to them. In the GEO environment upset rates may be as low as 10^{-10} errors/bit-day or as high as 10^{-4} errors/bit-day depending on the nature of the device in question. Upsets are directly dependent on the cross section, or LET, of the radiation for the device in question. This is sometimes called the *Heinrich flux*, as shown in figure 5.15. The more likely a device is to absorb radiation the more likely it is to experience an upset. Because SEEs are highly dependent on the specifics of both the radiation environment and the device in question the reader is referred to the references for more specific information.[28-41]

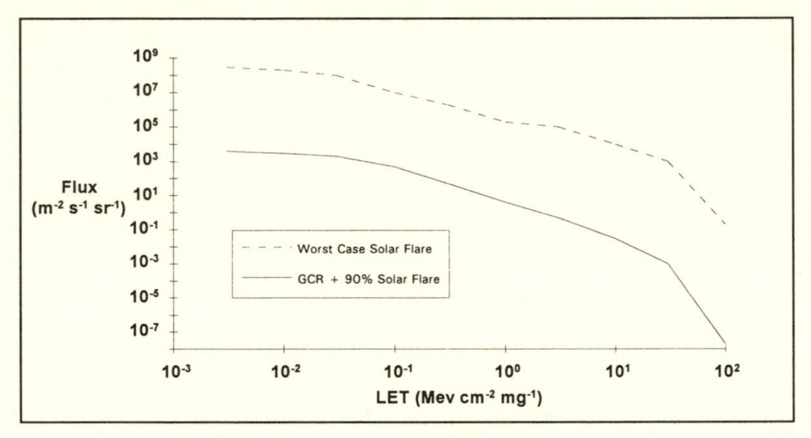

Fig. 5.15 Heinrich flux for a 25 mil spherical shell.

5.5 Models and Tools

Models of the trapped radiation environment are provided by the National Space Science Data Center. The current proton model is termed AP8, and is provided for both solar MIN and MAX conditions. The electron model is termed AE8. Both models are available in hard-copy format and software codes. The GCR and SPE models are detailed in the space station design documents, and are based on empirical fits to observations. A variety of radiation modeling codes are also available which can convert the anticipated spectrum of radiation into dose, for example, CREME (cosmic ray effects on microelectronics), and TREE (transient radiation effects on electronics).

5.6 Space and Ground-Based Testing

Because of the multitude of phenomena that may arise from the interaction of radiation with matter, radiation testing of devices and materials is an important aspect of validating spacecraft design. Key parameters that electronic devices are tested for include total dose effects, latchup threshold, upset threshold, single event upset, and neutron damage effects. A variety of equipment is available for use in these tests, with flash X ray machines and gamma immersion sources warranting further discussion.

Flash X ray machines are often used to simulate transient ionizing radiation. The device relies on a charged bank of capacitors for energy storage. When the capacitors are simultaneously discharged, the energy is used to generate an electron beam through the process of field emission. The electron beam alone may be used as a radiation source, or a tantalum plate can be used to convert the beam energy into brehmsstrahlung X rays. Doses on the order of 1 MRad (Si) can be obtained from the electron beam, or 1 KRad (Si) from the X rays, during operational times on the order of tens of nanoseconds. Dose and dose rate can be varied by adjusting the distance between the source and the object being irradiated.

Concentrated sources of radioactive materials can be used to subject samples to doses of either alpha, beta, or gamma radiation, depending on the nature of the source itself. Co^{60} and Cs^{137} are both gamma-emitting sources that are often used in immersion studies. Single event effects may be simulated with Cf^{252}, which emits alpha particles.

5.7 Design Guidelines

As shown in table 5.11, the designer has a variety of options to exercise in the radiation hardening of a spacecraft. On the part level, the key to radiation hardening is to select parts with sufficient margin. Parts with a factor of 5 safety margin may be used with confidence, parts with safety factors on the order of 2–5 may require additional testing and/or piece part traceability, whereas parts with safety factors of less than 2 should be avoided if possible. For single event upset, try to select devices with a LET threshold of > 100 $MeV/mg/cm^2$. On the system level, because certain failures and/or upsets are inevitable, the system must be capable of working around these events. Redundancy and software recovery algorithms are also critical to mission success.[9,42]

Table 5.11
Radiation Environment Effects Design Guidelines

Shielding	Place structures between sensitive electronics and the environment in order to minimize dose and dose rate
Design/Parts Selection	Utilize parts that have sufficient total dose safety factors and are latchup and upset resistant
Redundancy	Oversize solar arrays and design electronic systems with backup circuitry/parts
Recovery Algorithms	Install software capable of recovering the system from latchup or upsets

5.8 Exercises

1. The energy carried by a photon is related to its frequency by the relation $E = h\nu$. In general, an object's energy is related to its momentum and rest mass through the formula $E^2 = p^2c^2 + m^2c^4$. For particles at rest this reduces to the well known $E = mc^2$, and for photons it reduces to $E = pc$.

   Consider the interaction between a 1 g silicon semiconductor that is subjected to a 1 MRad dose of radiation in a time frame of 1 ms. Calculate

   a. The number of 10^{20} Hz gamma rays required to deliver the dose,

   b. The momentum pulse to the semiconductor,

   c. The force on the semiconductor, and

   d. The temperature rise if all of the absorbed energy appears as heat.

2. Estimate the shielding thickness of aluminum required to reduce the flux of 10^{20} Hz gamma rays by a factor of 2.

3. Calculate the daily radiation dose rate due to trapped electrons and protons at 1, 2, and 3 Earth radii for 1, 2, and 3 g/cm^2 shielding. Assume an equatorial orbit at 0 degrees inclination.

4. Consider the radiation emitted by a worst-case solar flare.

   a. What is the maximum proton energy that will be prevented from reaching a solar cell by a quartz coverslide that is 6 mils thick? 9 mils thick? (Assume a coverslide density of 2.5 g/cm^3.)

   b. If all of the radiation passing the coverslide is absorbed by the underlying cell, estimate the difference in radiation dose for both the 6 and 9 mil coverslides. Assume the radiation is deposited uniformly over a 24-hour time period.

5. A spacecraft is designed and successfully flown in an equatorial 500 km circular orbit. The customer wishes to fly an identical version of the spacecraft in a 20,000 km orbit. Discuss the differences between the radiation environments for the orbits and specify which interactions may be different.

6. Consider the Compton scattering of a photon, with initial energy E_γ, off of an atomic electron, which is assumed to be unbound and at rest. Assume that the photon is initially moving parallel to the horizontal. After the collision the electron is seen to be moving at an angle θ_e above the horizontal and the photon is seen to be moving at an angle θ_γ below the horizontal. Applying the principle's of conservation of energy and momentum, and remembering the relationships between energy and momentum discussed in exercise 1, derive an expression for the post collision energy of the photon as a function of θ_γ.

7. Section 5.2.2 discusses pair production, the process where an energetic photon spontaneously decays into an electron/positron pair.

   a. Show that if the only elements of the interaction are the photon, the electron, and the positron, it is impossible to simultaneously conserve energy and momentum. This is the confirmation of the fact that pair production can only occur near a nucleus.

   b. What is the miminum energy, and wavelength, a photon must have to be capable of initiating pair production?

5.9 Applicable Standards

ASTM E 170, *Standard Terminology Relating to Radiation Measurements and Dosimetry*, 25 May 1990.

ASTM E 512, *Standard Practice for Combined, Simulated Space Environment Testing of Thermal Control Materials with Electromagnetic and Particulate Radiation*, 27 December 1973.

ASTM E 665, *Standard Practice for Determining Absorbed Dose Versus Depth in Materials Exposed to the X Ray Output of Flash X Ray Machines*, 24 November 1978.

ASTM E 666, *Standard Practice for Calculating Absorbed Dose from Gamma or X Radiation*, 24 November 1978.

ASTM E 668, *Standard Practice for the Application of Thermoluminescence-Dosimetry (TLD) Systems for Determining Absorbed Dose in Radiation-Hardness Testing of Electronic Devices*, 29 December 1978.

ASTM E 763, *Standard Method for Calculation of Absorbed Dose from Neutron Irradiation by Application of Threshold-Foil Measurement Data*, 31 May 1985.

ASTM F 526, *Standard Test Method for Measuring Dose for use in Linear Accelerator Pulsed Radiation Effects Tests*, 29 May 1981.

ASTM F 867, *Standard Guide for Ionizing Radiation Effects Testing of Semiconductor Devices*, 27 May 1988.

ASTM F 1192, *Standard Guide for the Measurement of Single Event Phenomena (SEP) Induced by Heavy Ion Radiation of Semiconductor Devices*, 23 February 1990.

MIL-HDBK-279, *Total Dose Hardness Assurance Guidelines for Semiconductor Devices and Microcircuits*, 25 January 1985.

MIL-HDBK-280, *Neutron Hardness Assurance Guidelines for Semiconductor Devices and Microcircuits*, 19 February 1985.

5.10 References

1. Enge, H. A., *Introduction to Nuclear Physics* (Reading, MA: Addison-Wesley, 1966).
2. Bethe, H. A., and Heitler, W., "On the Stopping of Fast Particles and on the Creation of Positive Electrons," *Proc. Roy. Soc.* (London), A146, p. 83 (1934).
3. Jursa, A. S., ed., *Handbook of Geophysics and the Space Environment* (Air Force Geophysics Laboratory, Air Force Systems Command, United States Air Force, 1985).
4. Space Station Program Natural Environment Definiton for Design, *NASA Document SSP 30425, Rev. B* (Space Station Program Office, 1994).
5. Haffner, J. W., *Radiation and Shielding in Space* (San Francisco: Academic Press, 1967).
6. Gussenhoven, M. S., Brautigam, D. H., and Mullen, E. G., "Characterizing Solar Flare Particles in Near-Earth Orbits," *IEEE Tns. Nuc. Sci.*, 35, no. 6, pp. 1412–1419 (1988).
7. Rudie, N. J., *Principles and Techniques of Radiation Hardening*, vol.'s 1-12, 3d ed. (Hollywood, CA, Western Periodicals, 1986).
8. Langley, T. M., *Cassini Radiation Control Plan*, NASA JPL D-8813 (February 1993).
9. Tada, H. Y., Carter, J. R., Jr., Anspaugh, B. E., and Downing, R. G., *Solar Cell Radiation Handbook*, 3d ed., NASA JPL Publication 82-69 (1982).
10. Anon., *Solar Cell Array Design Handbook*, NASA JPL Publication SP 43-38 (1976).
11. Messenger, G. C., and Ash, M. S., *The Effects of Radiation on Electronic Systems*, 2d ed. (New York: Van Nostrand-Reinhold, 1992).
12. Ashcroft, N. W., and Mermin, N. D., *Solid State Physics* (Philadelphia: Holt, Rinehart and Winston, 1976).
13. Maurer, R. H., Herbert, G. A., and Kinnison, J. D., "Gallium Arsenide Solar Cell Radiation Damage Study," *IEEE Tns. Nuc. Sci.*, 36, no. 6, pp. 2083–2091 (1989).
14. Moreno, E. G., Alcubilia, R., Prat, L., and Castaner, L., "Radiation Damage Evaluation on AlGaAs/GaAs Solar Cells," *IEEE Tns. Nuc. Sci.*, 35, no. 4, pp. 1067–1071 (1988).
15. Laghari, J. R., and Hammoud, A. N., "A Brief Survey of Radiation Effects on Polymer Dielectrics," *IEEE Tns. Nuc. Sci.*, 37, no. 2, pp. 1076–1083 (1990).
16. Hammoud, A. N., Laghari, J. R., and Krishnakumar, B., "Electron Radiation Effects on the Electrical and Mechanical Properties of Polypropylene," *IEEE Tns. Nuc. Sci.*, 34, no. 6, pp. 1822–1826 (1987).

17. Long, S. A. T., Long, E. R., Jr., Ries, H. R., and Harries, W. L., "Electron-Radiation Effects on the AC and DC Electrical Properties and Unpaired Electron Densities of Three Aerospace Polymers," *IEEE Tns. Nuc. Sci.*, 33, no. 6, pp. 1390–1395 (1986).
18. Suthar, J. L., and Laghari, J. R., "Effect of Ionizing Radiation on the Electrical Characteristics of Polyvinylidene Fluoride (PVF_2)," *IEEE Tns. Nuc. Sci.*, 38, no. 1, pp. 16–19 (1991).
19. Glasstone, S., ed., *The Effects of Nuclear Weapons* (Washington DC: United States Atomic Energy Commission, 1962).
20. Nehru, J., ed., *Nuclear Explosions and Their Effects* (Publications Division, Ministry of Information & Broadcasting, Government of India, 1958).
21. Casarett, A. P., *Radiation Biology* (Englewood Cliffs, NJ: Prentice Hall, 1968).
22. Atwell, W., *Astronaut Exposure to Space Radiation*, SAE paper 901342, Intersociety Conference on Environmental Systems, Williamsburg, VA, 9–12 July 1990.
23. Atwell, W., Hardy, A. C., and Cash, B. L., "Space Shuttle Astronaut Exposures to the Space Radiation Environment: An Update," *NASA Technical Memorandum* (in press).
24. Donnelly, R. F., ed., *Solar-Terrestrial Predictions Proceedings,* vol. 2, *Working Group Reports and Reviews* (Boulder, CO: National Oceanic and Atmospheric Administration, 1979).
25. King, S. E., et al., "Radiation Survey of the LDEF Spacecraft," *IEEE Tns. Nuc. Sci.*, 38, no. 2, pp. 525–530 (1991).
26. Beever, E. R., and Rusling, D. H., "The Importance of Space Radiation Shielding Weight", *Second Symposium on Protection against Radiations in Space*, NASA SP-71 (1964).
27. Azarewicz, J. L., "Dose Rate Effects on Total Dose Damage," *IEEE Tns. Nuc. Sci.*, 33, no. 6, pp. 1420–1424 (1986).
28. Axness, C. L., Weaver, H. T., Fu, J. S., Koga, R., and Kolasinski, W. A., "Mechanisms Leading to Single Event Upset," *IEEE Tns. Nuc. Sci.*, 33, no. 6, p. 1577 (1986).
29. Binder, D., "Analytic SEU Rate Calculation Compared to Space Data," *IEEE Tns. Nuc. Sci.*, 35, no. 6, p. 1570 (1988).
30. Bion, T., and Bourrieau, J., "A Model for Proton-Induced SEU," *IEEE Tns. Nuc. Sci.*, 36, no. 6, p. 2281 (1989).
31. Adams, L., Harboe-Sorensen, R., Daly, E., and Ard, J., "Proton Induced Upsets in the Low Altitude Polar Orbit," *IEEE Tns. Nuc. Sci.*, 36, no. 6, p. 2339 (1989).
32. Bisgrove, J. M., Lynch, J. E., McNulty, P. J., Abdel-Kader, W. G., Kletnicks, V., and Kolasinski, W. A., "Comparison of Soft Errors

Induced by Heavy Ions and Protons," *IEEE Tns. Nuc. Sci.*, 33, n. 6, p. 1571 (1986).

33. Chlouber, D., O'Neill, P., and Pollock, J., "General Upper Bound on Single Event Upset Rate," *IEEE Tns. Nuc. Sci.*, 37, no 2, p. 1065 (1990).
34. Dyer, C. S., Sims, A. J., Farren, J., and Stephen, J., "Measurements of Solar Flare Enhancements to the Single Event Upset Environment in the Upper Atmosphere," *IEEE Tns. Nuc. Sci.*, 37, no. 6, p. 1929 (1990).
35. Dyer, C. S., Sims, A. J., Farren, J., and Stephen, J., "Measurements of the SEU Environment in the Upper Atmosphere," *IEEE Tns. Nuc. Sci.*, 36, no. 6, p. 2275 (1989).
36. Elder, J. H., Osborn, J., Kolasinski, W. A., and Koga, R., "A Method for Characterizing a Microprocessor's Vulnerability to SEU," *IEEE Tns. Nuc. Sci.*, 35, no. 6, p. 1678 (1988).
37. Harboe-Sorensen, R., Adams, L., and Sanderson, T. K., "A Summary of SEU Test Results Using Californium-252," *IEEE Tns. Nuc. Sci.*, 35, no. 6, p. 1622 (1988).
38. Harboe-Sorensen, R., Daly, E. J., Underwood, C. I., Ward, J., and Adams, L., "The Behavior of Measured SEU at Low Altitude during Periods of High Solar Activity," *IEEE Tns. Nuc. Sci.*, 37, no. 6, p. 1938 (1990).
39. Mullen, E. G., Gussenhoven, M. S., Lynch, K. A., and Brautigam, D. H., "DMSP Dosimetry Data: A Space Measurement and Mapping of Upset Causing Phenomena," *IEEE Tns. Nuc. Sci.*, 34, no. 6, p. 1251 (1987).
40. Normand, E., and Stapor, W. J., "Variation in Proton-Induced Upset Rates from Large Solar Flares Using an Improved SEU Model," *IEEE Tns. Nuc. Sci.*, 37, no. 6, p. 1947 (1990).
41. Srour, J. R., Hartmann, R. A., and Kitazaki, K. S., "Permanent Damage Produced by Single Proton Interactions in Silicon Devices," *IEEE Tns. Nuc. Sci.*, 33, no. 6, p. 1597 (1986).
42. Chen, C.-C., Liu, S.-C., Hsiao, C.-C., and Hwu, J.-G., "A Circuit Design for the Improvement of Radiation Hardness in CMOS Digital Circuits," *IEEE Tns. Nuc. Sci.*, 39, no. 2, p. 272 (1992).

6 The Micrometeoroid/Orbital Debris Environment

Then shall the dust return to the Earth as it was ...
—Ecclesiastes 12:7

6.1 Overview

On clear nights it is not unusual for a star gazer to witness the bright trails of a handful of shooting stars every hour. These shooting stars are actually tiny pieces of matter that burn up upon entering the Earth's atmosphere. Our solar system comes with its own naturally occurring background of dust that results from the breakup of comets, asteroids, and the like. These naturally occurring backgrounds are called *micrometeoroids* (MM). Since the launch of *Sputnik I*, mankind has been creating a new cloud of orbiting particles, which are the leftover pieces of nonoperational spacecraft, boost stages, solid rocket fuel particles, etc. The artificial environment is referred to as orbital debris (OD). Even though both MM and OD are, for the most part, quite small (< 1 cm), they are of great concern because of the large kinetic energies associated with impacts at orbital velocities.

6.2 Hypervelocity Impact Physics

Kinetic energy is given by the well-known relationship

$$KE = \frac{1}{2}mv^2. \tag{6.1}$$

Impacts on orbit occur at speeds on the order of 10 km/s. Assuming a density of 1 g/cm^3, particulate matter impacting at this speed will carry the kinetic energies shown in figure 6.1. Particles as small as 0.1 mm may give rise to surface erosion, while a 1 mm size particle may inflict serious damage. A 3 mm particle moving at 10 km/s carries the kinetic energy of a bowling ball moving at 100 km/hr, while a 1 cm particle has the kinetic energy of a 180 kg safe. These impacts can have potentially disastrous effects on a spacecraft (fig. 6.2). The characteristics of a hypervelocity impact are highly dependent upon impact velocity. At speeds less than about 2 km/s the impacting projectile will remain intact; between 2 and 7 km/s the particle will shatter into fragments; between 7 and 11 km/s the particle will go into a molten state; and above about 11 km/s the particle may vaporize. Depending on the state of the impacting particle, the physical processes responsible for transferring kinetic energy to the target may also vary. Along with the destruction of the impacting particle, some of the kinetic energy of impact will gouge out a crater in the target, as illustrated in figure 6.3.

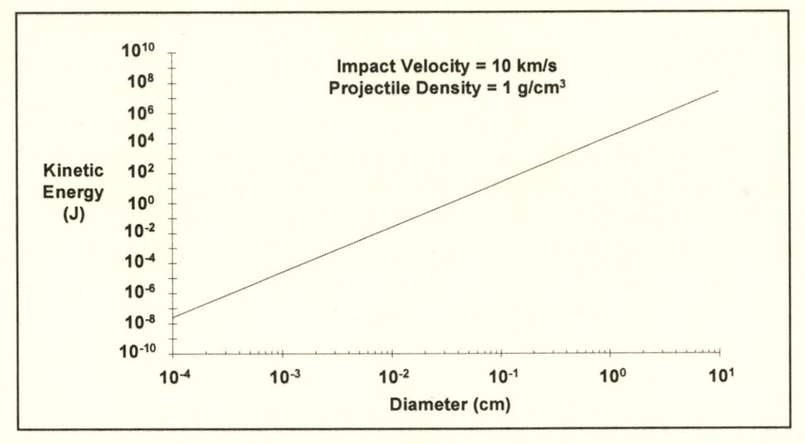

Fig. 6.1 Kinetic energy versus particle diameter.

The buildup of numerous craters on the surface of a material may greatly alter its surface properties. Similarly, if the crater is large enough the surface may be penetrated. This would allow the impacting material to damage components lying below the surface, compromising pressurized compartments or degrading structures and electronics. Numerous empirical observations indicate that the thickness of a material being penetrated, t (cm), is of the form

$$t \approx m_p^{\alpha/3} \rho_t^{\beta/3} v_\perp^{\gamma/3}, \tag{6.2}$$

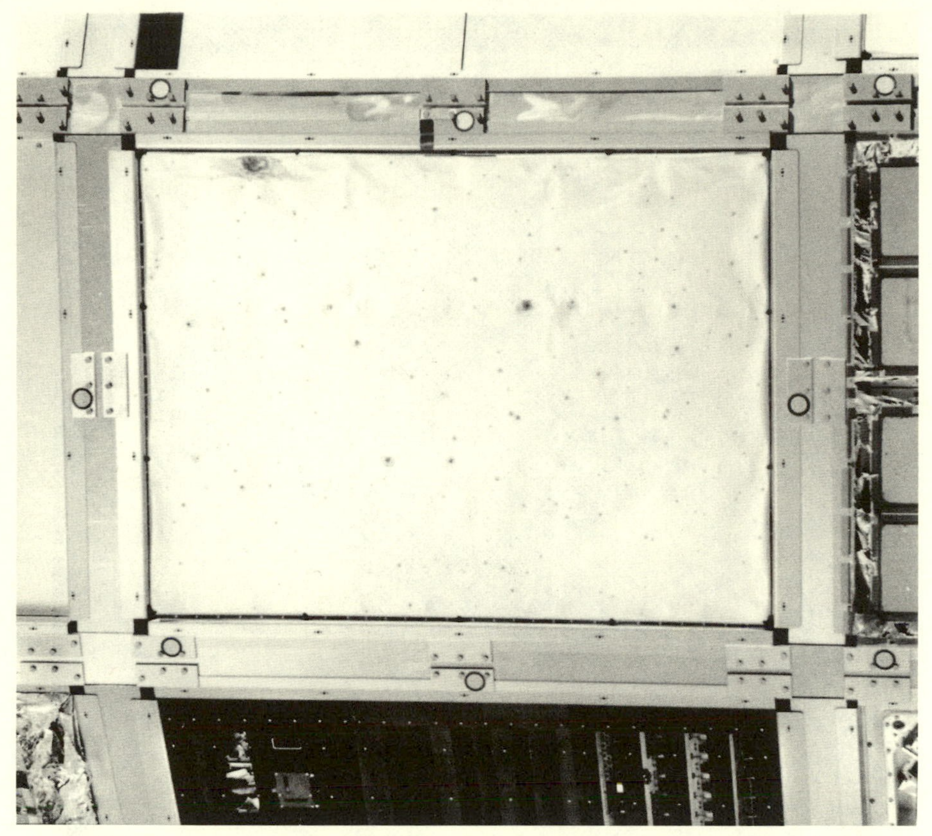

Fig. 6.2 Surface damage due to MMOD impacts on the Long Duration Exposure Facility (LDEF) during its 5 years and 9 months in LEO.
(Photograph courtesy of NASA)

where m_p (g) is the mass of the impacting projectile, ρ_t (g/cm^3) is the density of the target material, and $v_\perp$ (km/s) is the component of the impact velocity normal to the surface of the target.[1-6] The constants are generally on the order of $\alpha \sim 1$, $\beta \sim 0.5$, and $\gamma \sim 2$. Based on observations from the Long Duration Exposure Facility (LDEF), penetration thickness is approximated by

$$t = K_1 m_p^{0.352} \rho_t^{1/6} v_\perp^{0.875}, \tag{6.3}$$

where K_1 is a material constant.[7,8] LDEF results indicate that K_1 is equal to 0.72 for aluminum. Impacting projectiles which do not penetrate the surface will produce cratering. The crater depth, P (cm), is approximated by

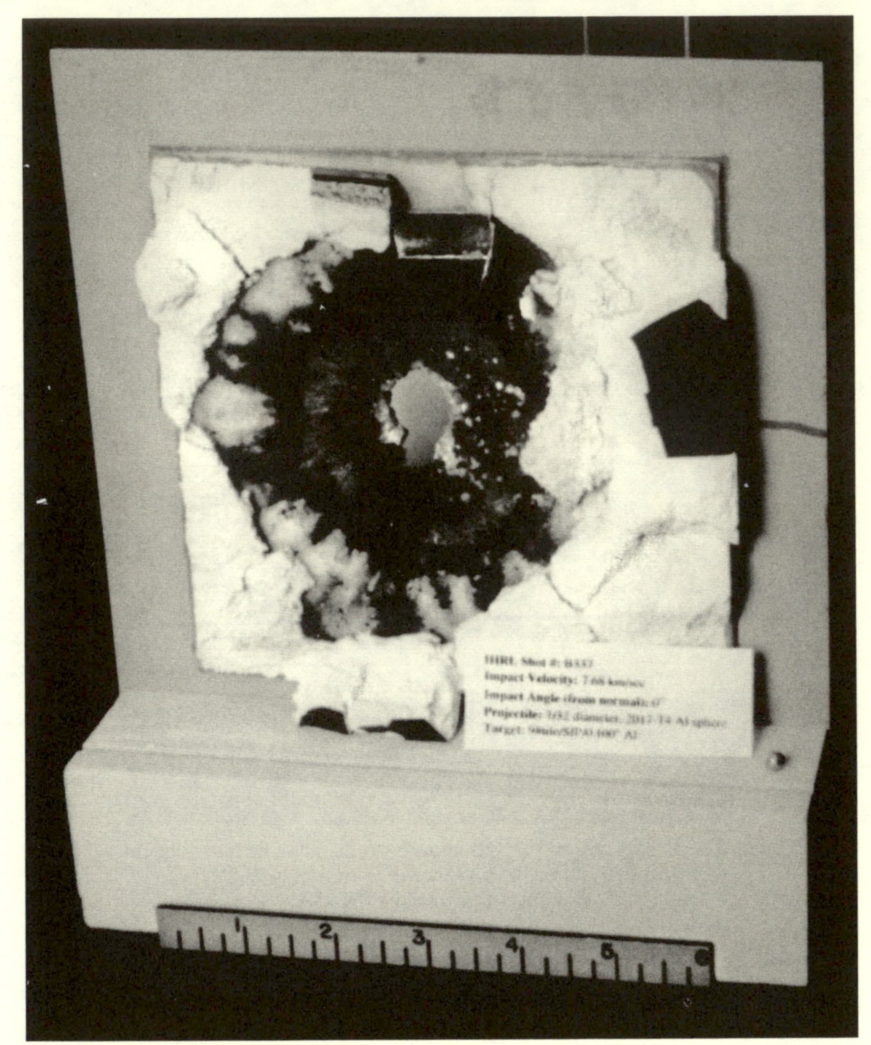

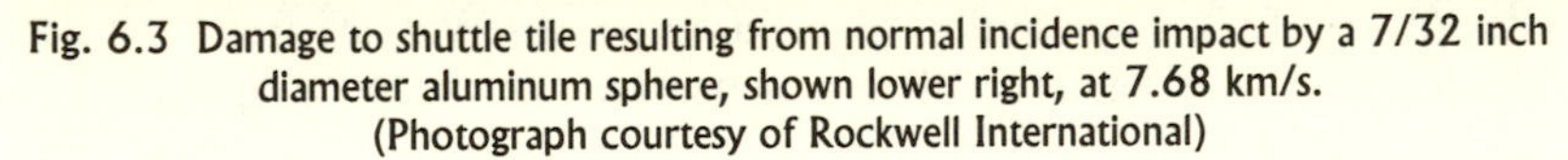
Fig. 6.3 Damage to shuttle tile resulting from normal incidence impact by a 7/32 inch diameter aluminum sphere, shown lower right, at 7.68 km/s. (Photograph courtesy of Rockwell International)

$$P = 0.42 m_p^{\,0.352} \rho_t^{\,1/6} v_{\perp}^{\,2/3} . \tag{6.4}$$

Crater depth and penetration thickness as a function of projectile size are illustrated in figure 6.4. Note that a projectile may alter surface properties near an impact site within an area bounded by 3 – 4 times the projectile diameter.

The primary cause of concern from both MM and OD is physical damage upon impact. Erosion of surface materials, changes in thermal control properties, the liberation of particles which may in turn contaminate sensitive surfaces, or the generation of EMI as particles vaporize upon impact are other possibilities. Although the EMI levels produced in these particle impacts are small, they may be detectable by sensitive spacecraft instrumentation. Electric field detectors on the *Voyager* spacecraft listened to the EMI associated with dust impacts on the antenna during the ring plane crossings as a means of inferring particle density during encounters with the outer planets.[9-11] Decompression of crew compartments on manned missions is also of concern. However, even if a projectile produced a pinhole-sized puncture of a *Salyut*-sized space station it would require almost a full day for the enclosed atmospheric pressure to dissipate. Even a pencil-sized hole would allow the crew about an hour to locate and repair the leak or don space suits.[12]

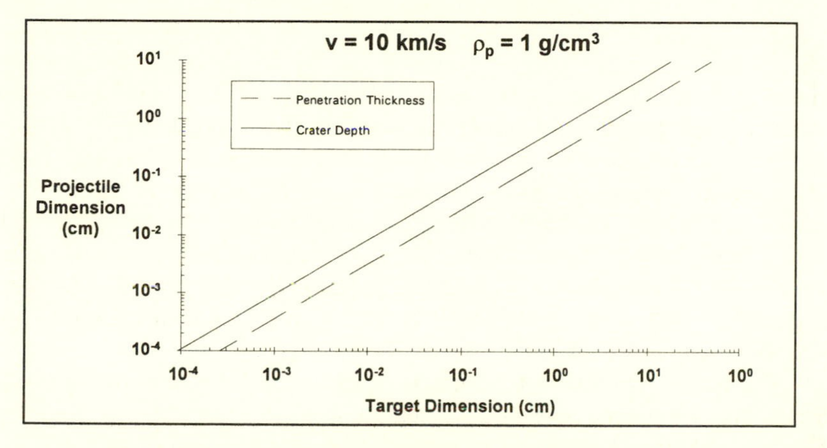

Fig. 6.4 Crater diameter and penetration thickness in aluminum.

6.3 The Micrometeoroid Environment

Observations of larger MMs with ground-based radar, and examination of surfaces exposed to hypervelocity impacts on orbit, allow researchers to infer the size and frequency distribution of the naturally occurring MM environment. The MM flux is not constant, but varies slightly over the course of the year during events known as *meteor showers*.[13] These are times when the Earth's orbit intersects the orbital path of the cloud of dust left by the breakup of a comet, for example. In 1993 the launch of the space shuttle was delayed for one day to allow the Persied meteor shower, which always peaks around August 11, to decrease. However, since most spacecraft have orbital lifetimes on the order of months if not years, an average background value is sufficient for most purposes. The interplanetary MM background is described by the relation

$$F_{MM}\left(m^{-2}yr^{-1}\right) = 3.156\times10^{7}\left[A^{-4.38}+B+C\right] \tag{6.5}$$

where

$$A = 15+2.2\times10^{3}m^{0.306}$$

$$B = 1.3\times10^{-9}\left(m+10^{11}m^{2}+10^{27}m^{4}\right)^{-0.306}$$

$$C = 1.3\times10^{-16}\left(m+10^{6}m^{2}\right)^{-0.85}$$

and m (g) is the MM mass. This is illustrated in figure 6.5. The chances of impacting a larger piece of matter are quite small, but impacts with smaller pieces are certainties on longer missions. The average MM velocity is 17 km/s, which corresponds to an average impact velocity of 19 km/s. The normalized velocity distribution is approximated by 0.112 for the range 11.1 – 16.3 km/s, by $[(3.328\times10^{5})v^{-5.34}]$ for the range 16.3 – 55 km/s, and by 1.695×10^{-4} for the range 55 – 72.2 km/s. Recommended mass densities are 2 g/cm^3 for MM < 10^{-6} g, 1 g/cm^3 for 10^{-6} g to 0.01 g, and 0.5 g/cm^3 for masses greater than 0.01 g.[14]

The Earth's gravitational field acts to focus the MMs toward the Earth and at the same time shields a spacecraft from impacts coming from the direction of the Earth. To account for gravitational focusing, the interplanetary flux is multiplied by the factor

$$F_{grav} = 1+\frac{R_E+100km}{R_E+h}, \tag{6.6}$$

where h (km) is the spacecraft altitude. Because the Earth will also intercept any MM coming at the spacecraft from the direction of the Earth (fig. 6.6), the interplanetary flux is reduced by the Earth shielding factor

$$F_{shield} = \frac{1+\cos\eta}{2}, \tag{6.7}$$

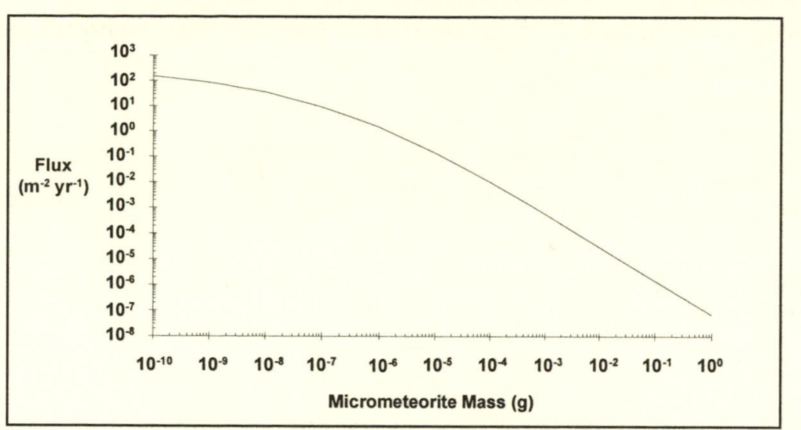

Fig. 6.5 The interplanetary micrometeoroid flux.

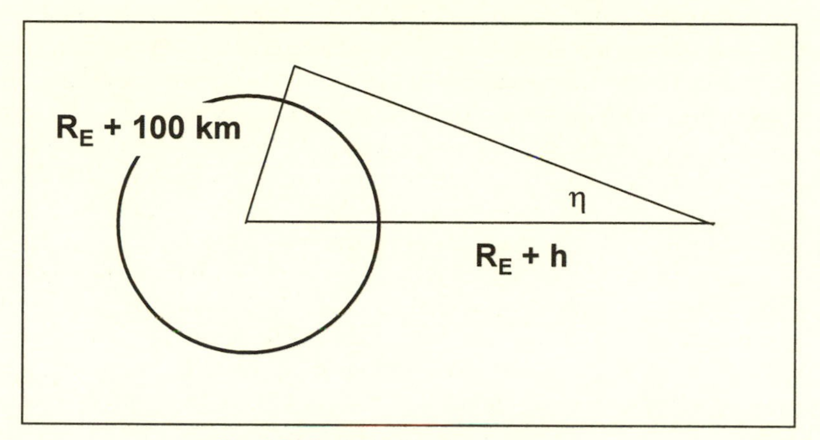

Fig. 6.6 Earth shielding factor geometry.

where

$$\eta = \sin^{-1}\left(\frac{R_E + 100km}{R_E + h}\right). \tag{6.8}$$

Note that equation (6.7) implicitly assumes that any piece of matter passing within 100 km of the Earth's surface will reenter. This is in agreement with the assumption made in chapter 1 that space starts at 100 km altitude. Finally, because of the nature of the distribution, most impacts can be expected on space-facing surfaces. Side-facing or Earth-facing surfaces would expect to see a reduced flux. For Earth-facing or trailing surfaces the flux is reduced by a factor of 10, while ram or side-facing surfaces see a reduction by a factor

$$F_{dir} = \frac{1.8 + 3\sqrt{1 - \left(\frac{R_E + 100km}{R_E + h}\right)^2}}{4}. \tag{6.9}$$

The flux to any particular location on a spacecraft is the product of equations (6.5), (6.6), (6.7), and (6.9).

6.4 The Orbital Debris Environment

As the name implies, artificial OD is found in orbit around the Earth with velocities on the order of 8 km/s. Consequently, OD will impact a spacecraft at lower velocities than will MMs and OD collisions will occur on mainly ram-facing surfaces. Unlike MM, the OD flux is affected by the solar cycle (via aerodynamic drag). More importantly, if one compares fluxes of equal size particles, the OD environment in some of the more popular orbits currently exceeds the MM environment. There may be as many as 20,000 pieces of debris > 4 cm now circling the Earth.[12,15] Most OD is < 1 cm in size, but only those objects greater than 10 cm in size (~ 7000 in number) are tracked routinely (fig. 6.7). The total mass of OD continues to grow as the satellite population increases at the rate of about 230 objects per year.[12,16]

Sources of OD include nonoperational spacecraft, boost vehicles, spacecraft explosions (when propellant becomes unstable at end of life) or breakups, collisions, solid rocket fuel particulates, and surface erosion particulates. The material that is ejected from the surface of a material during a hypervelocity impact may go on to become orbital debris itself. As shown in table 6.1, 1 kg of aluminum may be used to form several hundred thousand 1 mm size particles. Even small masses may contribute significantly to the OD problem.

Although the OD environment is not well defined or understood, and the number of small (< 1 cm) OD above 700 km is virtually unknown, the OD flux, in units of particles of diameter d (cm) per m^2 per year, can be approximated by

$$F_{OD}\left(m^{-2}yr^{-1}\right) = H(d)\phi(h,S)\Psi(i) \times\left[F_1(d)g_1(t) + F_2(d)g_2(t)\right] \tag{6.10}$$

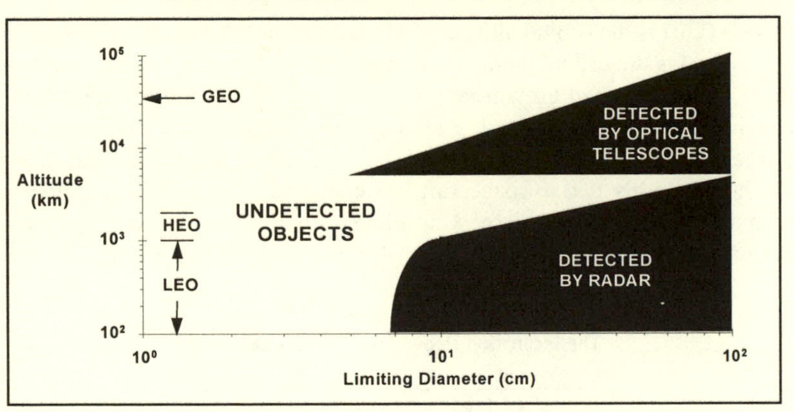

Fig. 6.7 Detectable size of objects in orbit.

Table 6.1
Number of Particles of Diameter d per kg of Aluminum

Particle Diameter, *d* (cm)	0.1	0.2	0.5	1.0
# of Particles of Dia. *d*/kg Al	707,617	88,452	5661	707

with

$$H(d) = \left\{ 10^{\exp\left[-\left(\frac{\log d - 0.78}{0.637}\right)^2\right]} \right\}^{1/2}$$

$$\phi(h,S) = \frac{\phi_1(h,S)}{\phi_1(h,S)+1} \qquad \phi_1(h,S) = 10^{\left[\frac{h}{200} - \frac{S}{140} - 1.5\right]}$$

$$\mathrm{F_1(d)} = \left(1.22 \times 10^{-5}\right)\mathrm{d}^{-2.5}$$

$$F_2(d) = \left(8.1 \times 10^{10}\right)(d + 700)^{-6}$$

$$g_1(t) = (1+q)^{(t-1988)}$$

$$g_2(t) = 1 + p(t - 1988),$$

where h (km) is the orbital altitude, S (F10.7) is the 13-month smoothed solar flux, i (deg) is the orbital inclination, t (year < 2011) is the year in question, $p \sim 0.05$ is the assumed growth rate of intact objects, $q \sim 0.02$ is the estimated growth rate of fragments, and $\Psi(i)$ is a function describing the relationship between flux and inclination.[14,17-19] $\Psi(i)$ is given in table 6.2. Because the OD flux is closely tied to spacecraft launches, the most popular altitudes and inclinations are where the largest problems are expected. The projected OD environment for typical LEO orbit in 1995 is shown in figure 6.8.

Table 6.2
The Inclination Dependence of Orbital Debris

i (deg)	$\Psi(i)$
28.5	0.91
30	0.92
40	0.96
50	1.02
60	1.09
70	1.26
80	1.71
90	1.37
100	1.78
120	1.18

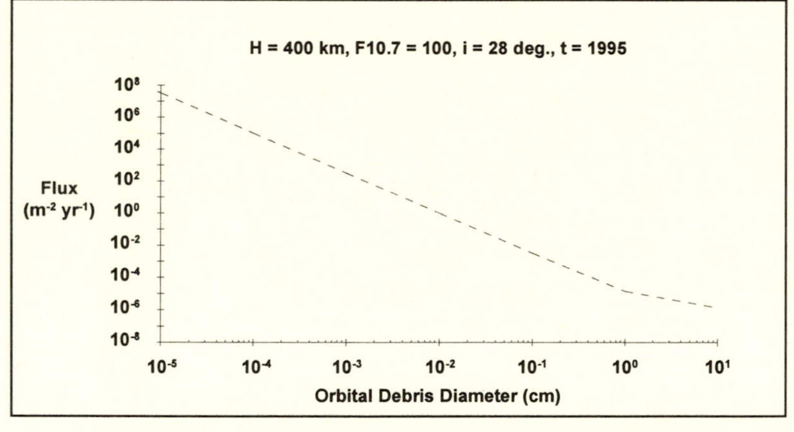

Fig. 6.8 The orbital debris environment for a typical LEO orbit.

The mass density for debris smaller than 0.62 cm is ~ 4 g/cm^3, the mass density for debris larger than 0.62 cm is ~ $2.8d^{-0.74}$ g/cm^3, where d (cm) is the particle diameter. As previously mentioned, the OD flux now exceeds the MM flux for certain orbits (fig. 6.9). Because aerodynamic drag can help to clear the OD environment at the lower altitudes, the OD flux is seen to increase logarithmically between about 300–1000 km and remains fairly constant in the 1000–2000 km range.[20] As a quantitative measure of the growing significance of the OD problem, during the first 30 shuttle missions there had been damage to 27 windows on 18 flights. On STS-7 a 0.2 mm diameter fleck of paint impacted a side window, which required replacement at a cost of $50,000 upon completion of the mission.

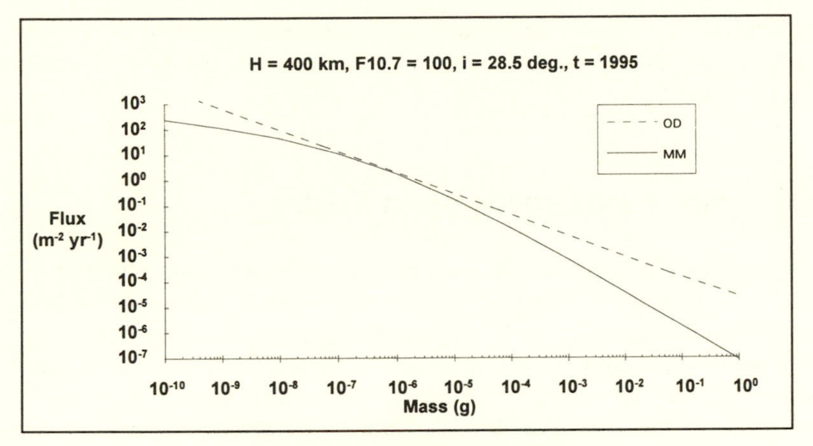

Fig. 6.9 The MMOD flux in LEO.

The issue of OD is further complicated by the mechanics of the evolution of the debris cloud following the breakup of an orbiting object.[21,22] If a propulsion tank were to explode, for example, the ejected pieces of matter would spread out from the point of origin with slightly different initial conditions. Depending on the orbital conditions at the point of breakup, and the nature of the breakup itself, the pieces may remain fairly well correlated for some time or may rapidly disperse into dissimilar orbits. Oftentimes, the most likely point of impact on a spacecraft is not the ram surface but the sides.

6.5 Impact Probabilities

The probability of an impact with an object can be estimated from elementary statistics. In time T (years) the number of impacts an object of surface area A (m^2) can expect is given by

$$N = \int_{t}^{t+T} FAdt, \tag{6.11}$$

while the probability of n impacts is given by

$$P_n = \frac{N^n}{n!} \exp^{-N}. \tag{6.12}$$

To estimate the number of impacts due to objects greater than a specific mass, simply exclude the lower mass particles from the flux in equation (6.11).

6.6 Space and Ground-Based Testing

Studies of hypervelocity impacts on the ground are hampered by the fact that it is difficult to accelerate larger mass objects to orbital velocities. Masses in the range 10^{-15} to 10^{-10} g have been charged and accelerated up to 40 km/s with the aid of a Van de Graff accelerator.[1] Larger masses, on the order of g, must be studied with light gas guns which accelerate the particles by means of high pressure flow.[23] Sandia Lab has accelerated 1 g class projectiles to 16 km/s. Because the collision happens so rapidly it is not possible to monitor the propagation of the shock wave through the material in question. These studies must be performed numerically.

6.7 Design Guidelines

Spacecraft thermal blankets or structural panels can easily shield the smaller MMOD (< 0.1 mm). Because of the nature of the MMOD environment, there are few active steps that can be taken to minimize the possibility, or consequences, of larger MMOD impacts. As shown in table 6.3, orienting sensitive surfaces away from ram or flying at altitudes/inclinations that minimize OD are about the only options. On missions where collisions with larger MMOD are expected, spacecraft designers may utilize a "bumper" on the leading edge of the spacecraft.[24] A "bumper" would consist of 2 sheets of material separated by a few centimeters. The outer layer would be penetrated by the MMOD impact, but would cause the projectile to fragment into smaller pieces of matter, which could then be stopped by the back layer of the bumper. Stiffening the bumper with Kevlar and Nextel layers can reduce the number of miniprojectiles created by the initial impact. Because this configuration adds mass and volume to the spacecraft, it is not an option that can be exercised by many programs.

Table 6.3
MMOD Environment Effects Design Guidelines

Orientation	Point sensitive surfaces away from ram
Orbit	Fly at altitudes/inclinations where debris is minimized
Shielding	Utilize leading edge bumper shield to protect critical components

Of equal importance in design of spacecraft are methods to minimize the production of OD that could cause problems on future missions. In order to control the growth of OD, a requirement on most new spacecraft programs is to minimize the amount of orbital debris produced by the mission. The United States issued a policy statement on the subject in 1988: "All space sectors will seek to minimize the creation of space debris. Design and operations of space tests, experiments and systems will strive to minimize or reduce accumulation of space debris consistent with mission requirements and cost-effectiveness. The United States government will encourage other space faring nations to adopt policies and practices aimed at debris minimization."[25] Even the Kinetic Energy Anti-Satellite (KE-ASAT) program, which was designed to destroy satellites with the physical force of an impact at orbital speeds, had a requirement to perform this task using a deployable paddle as the impacting object, rather than the entire satellite, so as to minimize the amount of OD produced by the collision. Many methods exist for minimizing debris production, such as avoiding single stage to orbit boosters (which then remain in orbit), using longer life satellites, venting fuel tanks at the end of a mission (to minimize the possibility of explosion), and moving inactive satellites to either lower orbits with faster decay times or to unused orbits.

6.8 Exercises

1. From the definition, $d\omega = \sin\theta \, d\theta \, d\phi$, calculate the solid angle that the Earth subtends as seen by a satellite orbiting at altitude h (fig. 6.6).

2. Bearing in mind that the Earth shielding factor must have the value 1 at $h = \infty$ (interplanetary space) and the value 0.5 at h = 100 km, (the top of the Earth's atmosphere), show that the expression for the solid angle found in exercise 1 can be used to reproduce the correct expression for Earth shielding factor (equation 6.7).

3. Consider the MMOD environment in a 400 km, 56 degree inclination orbit for the year 1998. Infer the solar cycle conditions from the data provided in the text.

    a. Plot the flux of both MM and OD as a function of particle size.

    b. What size of MMOD would have a "noticeable" effect on a solar array? Define "noticeable" and assume a 6 mil, silicon coverslide.

    c. What is the likelihood of an MMOD impact that "noticeably" damages a 5 m^2 solar array?

    d. Should the array be oversized to allow for degradation by MMOD impacts? Why or why not?

6.9 References

1. Dietzel, H., Neukum, G., and Rauser, P., "Micrometeoroid Simulation Studies on Metal Targets," *J. Geophys. Res.*, 77, no. 8, p. 1375 (1972).
2. Titov, V. M., and Fadeenko, Yu. I., "Puncturing of Spacecraft Hulls by Meteoric Impact," *Kosmicheskie Issledovaniya*, 10, no. 4, pp. 598-595 (1972).
3. Lawrence, R. J., "Enhanced Momentum Transfer from Hypervelocity Particle Impacts," U.S. Department of Energy, Sandia National Laboratories, Contract DE-AC04-76DP00789.
4. Pailer, N., and Grun, E., "The Penetration Limit of Thin Films," *Planet. Space Sci.*, 28, pp. 321-331 (1980).
5. McDonnell, J.A.M., ed., *Cosmic Dust* (New York: John Wiley & Sons, 1978).
6. Zukas, J. A., Nicholas, T., Swift, H. F., Greszczuk, L. B., and Curran, D. R., *Impact Dynamics* (New York: John Wiley & Sons, 1982).
7. *LDEF — 69 Months in Space: First Post-Retrieval Symposium*, parts 1–3, NASA CP-3134 (1991).
8. *LDEF — 69 Months in Space: Second Post-Retrieval Symposium*, parts 1–4, NASA CP-3194 (1992).

9. Gurnett, D. A., Grun, E., Gallagher, D., Kurth, W. S., and Scarf, F. L., "Micron-Sized Particles Detected near Saturn by the Voyager Plasma Wave Instrument," *Icarus*, 53, pp. 236–254 (1983).
10. Gurnett, D. A., Kurth, W. S., Scarf, F. L., Burns, J. A., Cuzzi, J. N., and Grun, E., "Micron-Sized Particles Detected near Uranus by the Voyager 2 Plasma Wave Instrument," *J. Geophys. Res.*, 92, no. A13, pp. 14,959–14,968 (1987).
11. Gurnett, D. A., Kurth, W. S., Granroth, L. J., Allendorf, S. C., and Poynter, R. L., "Micron-Sized Particles Detected near Neptune by the Voyager 2 Plasma Wave Instrument," *J. Geophys. Res.*, 96, suppl., pp. 19,177–19,186 (1991).
12. Johnson, N. L., and McKnight, D. S., *Artificial Space Debris* (Malibar, FL: Orbit Book Co., 1987).
13. Baggaley, W. J., and Steel, D. I., "The Seasonal Structure of Ionosonde Es Parameters and Meteoroid Deposition Rates," *Planet. Space Sci.*, 32, no. 12, pp. 1533–1539 (1984).
14. *Space Station Program Natural Environment Definition for Design*, Space Station Program Office, SSP 30425, Rev. B (1994).
15. *Report on Orbital Debris*, Interagency Group (Space), National Security Council (February 1989).
16. Taft, L. G., "Satellite Debris: Recent Measurements," *J. Spacecraft*, 23, no. 3, p. 342 (1986).
17. Reynolds, R. C., Fischer, N. H., and Rice, E. E., "Man-Made Debris in Low Earth Orbit — A Threat to Future Space Operations," *J. Spacecraft*, 20, no. 3, p. 279 (1983).
18. Kessler, D. J., "Sources of Orbital Debris and the Projected Environment for Future Spacecraft," *J. Spacecraft*, 18, no. 4, p. 357 (1981).
19. Kessler, D. J., and Anz-Meador, P. D., "Effects on the Orbital Debris Environment due to Solar Activity," paper 90-0083, American Institute of Aeronautics and Astronautics, 29th Aerospace Sciences Meeting, Reno, NV (1990).
20. Chobotov, V., and Spencer, D., "Debris Evolution and Lifetime Following an Orbital Breakup," paper 89-0085, American Institute of Aeronautics and Astronautics, 28th Aerospace Sciences Meeting, Reno, NV (1989).
21. McKnight, D., and Brechin, C., "Debris Creation via Hypervelocity Impact," paper 90-0084, American Institute of Aeronautics and Astronautics, 29th Aerospace Sciences Meeting, Reno, NV (1990).
22. Mog, R., "Spacecraft Protective Structures Design Optimization," paper 90-0087, American Institute of Aeronautics and Astronautics, 29th Aerospace Sciences Meeting, Reno, NV (1990).
23. Taylor, R. A., "A Space Debris Simulation Facility for Spacecraft Materials Evaluation," *SAMPE Quarterly*, 18, no. 2, pp. 28-34 (1987).

24. Humes, D. A., "Hypervelocity Impact Tests on Space Shuttle Orbiter RCC Thermal Protection Material," *J. Spacecraft*, 15, no. 4, p. 250 (1978).
25. Spralding, K., "Space Debris: The Legal Regime, Policy Considerations and Current Initiatives," paper 90-0088, American Institute of Aeronautics and Astronautics, 29th Aerospace Sciences Meeting, Reno, NV (1990).

7 Conclusions

Better is the end of a thing than the beginning thereof.
—Ecclesiastes 7:8

7.1 Overview

As has been seen in the preceding chapters, the space environment may have a direct impact on a spacecraft subsystem's ability to execute its design objective. Depending on the severity of the orbit, these interactions may be quite mild or may be mission threatening (fig. 7.1).[1,2] Spacecraft subsystems must often carry extra margin in their designs in order to allow for the interactions previously discussed. Although each of these interactions may be quite severe in their own right, they are not the entire story. As illustrated in figure 7.2, there is the very real possibility for synergistic interactions between the various environments which could result in total degradation that is worse than the sum of its parts. Oftentimes the environments found on the vertical axis can increase the degradations listed on the horizontal axis. Experience on the part of the spacecraft designer, combined with the level of risk deemed acceptable by the spacecraft program, are the best tools to use in estimating the potential for synergistic interactions.

7.2 References

1. Tribble, A. C., "The Space Environment and its Impact on Spacecraft Design," paper 93-0491, American Institute of Aeronautics and Astronautics, 32d Aerospace Sciences Meeting, Reno, NV (1993).
2. Tribble, A. C., "Spacecraft Interactions with the Space Environment," paper 95-0838, American Institute of Aeronautics and Astronautics, 34th Aerospace Sciences Meeting, Reno, NV (1995).

	SPACE ENVIRONMENTS										
	VACUUM		NEUTRAL				PLASMA	RADIATION			MMOD
SPACECRAFT SYSTEMS	Solar UV	Outgassing/ Contamination	Aerodynamic Drag	Sputtering	Atomic Oxygen Attack	Spacecraft Glow	Spacecraft Charging	Van Allen Belts	Galactic Cosmic Rays	Solar Proton Events	Impacts
Attitude Determination & Control		Degradation of Sensors	Induced Torques	Change in sensor coatings		Interference with sensors	Torques due to induced potentials				
Avionics							Upsets due to EMI from arcing	Degradation: SEU's, bit errors, ...			EMI due to impacts
Electrical Power	Change in coverslide transmittance			Change in coverslide transmittance			Shift in floating potential, reattraction of contaminants	Degradation of solar cell output			Destruction/ Obscuration of solar cells
Propulsion		Thruster plumes may be a contaminant source	Drag Makeup Fuel Requirement				Shift in floating potential due to thruster firings				Rupture of pressurized tanks
Structures											Penetration
Telemetry, Tracking, & Communications		Degradation of Sensors		Change in sensor coatings		Interference with sensors	EMI due to arcing	Degradation of electronics			EMI due to impacts
Thermal Control	Change in surface alpha/epsilon ratio			Change in surface alpha/epsilon ratio			Reattraction of contaminants	Cold surfaces may experience heating			Degradation of alpha/epsilon

Fig. 7.1 Space environment effects.

		VACUUM		NEUTRAL				PLASMA	RADIATION			MMOD
		Solar UV	Outgassing/ Contamination	Aerodynamic Drag	Sputtering	Atomic Oxygen Attack	Spacecraft Glow	Spacecraft Charging	Van Allen Belts	Galactic Cosmic Rays	Solar Proton Events	Impacts
VACUUM	Solar UV		Photochemical Deposition of Contaminants	Solar cycle alters atmospheric densities				Induces photoemission of electrons				Solar cycle alters OD density
	Outgassing/ Contamination						Outgassed material may contribute to glow	Increases arcing rate				
NEUTRAL	Aerodynamic Drag		Flow may reflect contaminants to S/C									Drag removes OD from lower orbits
	Sputtering		Sputtered material may contaminate sensitive surfaces									
	Atomic Oxygen Attack		AO may clean contaminated surfaces				AO resistant materials are susceptible to glow	AO attack may alter surface conductivities				
	Spacecraft Glow											
PLASMA	Spacecraft Charging				Charged surfaces may increase sputtering							
RADIATION	Van Allen Belts		Radiation dose may increase outgassing					Radiation may increase charging				
	Galactic Cosmic Rays											
	Solar Proton Events									SPEs suppress GCRs		
MM/OD	Impacts		Impacts may liberate contaminants	Impacts may slightly increase drag		Impacts may expose underlying surfaces to erosion		Impact vaporization may stimulate arcing				

Fig. 7.2 Synergistic space environment effects.

Appendix 1

Nomenclature

Symbols

a = acceleration; impact parameter
A = area
B = brightness; magnetic field
c = speed of light
C = capacitance; coefficient; constant
d = diameter, distance
e = elementary charge
E = electric field; emissive power; energy; irradiance
f = frequency; function
F = force; solar cell output; radiation view factor
g = acceleration due to gravity; carrier generation constant; number of sunspot groups
G = gravitational constant
h = height (altitude); Planck constant
H = scale height
I = current; impulse; intensity; response
J = current density
k = Boltzmann constant
K = magnetic index
l = length
L = length; irradiance; constant
m = mass
M = magnetic moment
n = number density
N = number density; number of impacts; numerical designation
p = momentum impulse; constant
P = pressure; probability
q = charge; constant
Q = charge; constant; heat
r = radius
R = gas constant; range; reflectance; solar activity; radius
s = number of individual sunspots
S = solar flux
t = time; penetration thickness
T = kinetic energy; period; temperature
U = binding energy; potential
v = velocity
V = potential; volume
W = width
x = size; thickness
X = accumulation rate
Y = yield
Z = atomic number

α = absorptance; accommodation
δ = distance
Δ = change
ε = emittance; permitivity
ϕ = angle; flux; impact energy
γ = sticking coefficient; ratio
λ = wavelength; Debye length

μ = cross section; magnetic moment; absorption coefficient
θ = angle
ρ = mass density; charge density
σ = cross section; momentum accommodation; Stefan-Boltzmann constant
τ = residence time
ω = solid angle
ξ = "thickness"

Subscripts

a = activation; array
c = contamination; cyclotron
d = drag; drift
D = drag
e = electron
E = Earth
f = fuel; final
fl = floating
i = ion; initial; impact; incident
ls = lateral surface
L = loss
n = normal
o = orbital; initial; plasma
p = plasma; planetary
r = reflected; radial
s = solar; station
s/c = spacecraft
sp = specific
$stop$ = stopping
t = tangential; target
th = thermal; threshold
tot = total
w = accommodation

Appendix 2

Acronyms

ADC = attitude determination and control
AO = atomic oxygen
BRDF = bidirectional reflectance distribution function
CME = coronal mass ejection
EMI = electromagnetic interference
EPS = electrical power system
EWB = Environmental Workbench
ESD = electrostatic discharge
GCR = galactic cosmic ray
GEO = Geosynchronous
HEO = high Earth orbit; highly elliptical orbit
IGRF = international geomagnetic reference field
IO = indium oxide
ISO = international standards organization
ITO = indium tin oxide
LDEF = Long Duration Exposure Facility
LEO = low Earth orbit
LET = linear energy transfer
MM = micrometeoroid
MSIS = mass spectrometer incoherent scatter
NASA = National Aeronautics and Space Administration
NASCAP = NASA charging analysis program
OD = orbital debris
PST = point source transmittance
RBE = relative biological effectiveness
RE = reaction efficiency
REM = Roentgen equivalent in man
RER = reaction efficiency ratio
RHU = radioisotope heating unit
RTG = radioisotope thermoelectric generator
SEE = single event effect; space environment effect
SEP = single event phenomena
SEU = single event upset
SNR = signal to noise ratio
SPE = solar proton event
TCS = thermal control system
TT&C = telemetry, tracking & communication
UV = ultraviolet
VAB = Van Allen belt
QCM = quartz crystal microbalance

■ Appendix 3

Physical Constants

Fundamental Constants

Constant	Symbol	Value
Speed of light	c	3.00×10^{8} m/s
Elementary charge	e	1.60×10^{-19} C
Planck constant	h	6.63×10^{-34} Js
Gravitational constant	G	6.67×10^{-11} m^3/s^2 kg
Boltzmann constant	k	1.38×10^{-23} J/K
Electron rest mass	m_e	9.11×10^{-31} kg
Neutron rest mass	m_p	1.67×10^{-27} kg
Proton rest mass	m_n	1.68×10^{-27} kg
Avogadro's number	N_A	6.02×10^{23} mol^{-1}
Gas constant	R	1.985×10^{-3} kcal/mol K
Permittivity constant	ε_0	8.85×10^{-12} F/m
Stefan-Boltzmann constant	σ	5.67×10^{-8} W/m^{2}K^4

Astronomical Data

Object	Property	Value
Sun	Mass	1.99×10^{30} kg
	Mean radius	6.96×10^{8} m
	Total energy output	3.90×10^{26} W
	Energy density at Earth	1340 W/m^2
Earth	Mass	5.98×10^{24} kg
	Mean radius	6.40×10^{6} m
	Distance from Sun	1.50×10^{11} m

Appendix 4

The Long Duration Exposure Facility

NASA's Long Duration Exposure Facility (LDEF) was a free-flying, twelve-sided cylindrical spacecraft, measuring 30 feet in length and 14 feet in diameter, that was designed to test the stability of materials and systems to exposure in the LEO environment (fig. A4-1). The LDEF was three-axis stabilized, to insure highly reliable predictions of environmental exposure conditions, and carried fifty-seven separate experiments in areas such as materials, coatings, thermal systems, power, propulsion, space science, electronics, and optics.[1] Most of the experiments were passive, with the majority of the data resulting from postflight analysis. The LDEF was placed in LEO by the space shuttle *Challenger* in April 1984, with the intention of remaining in orbit for one full year until captured and retrieved on a later mission. Before the retrieval could occur, the shuttle fleet was grounded as the result of the *Challenger* accident, and it was 5 years and 9 months before the spacecraft was returned in January 1990 by the shuttle *Columbia*. The extended mission resulted in a wealth of data on the interaction of the experiments with the LEO environment which have been presented at three dedicated postretrieval symposiums and integrated into the Materials and Processes Technical Information Service (MAPTIS) database.

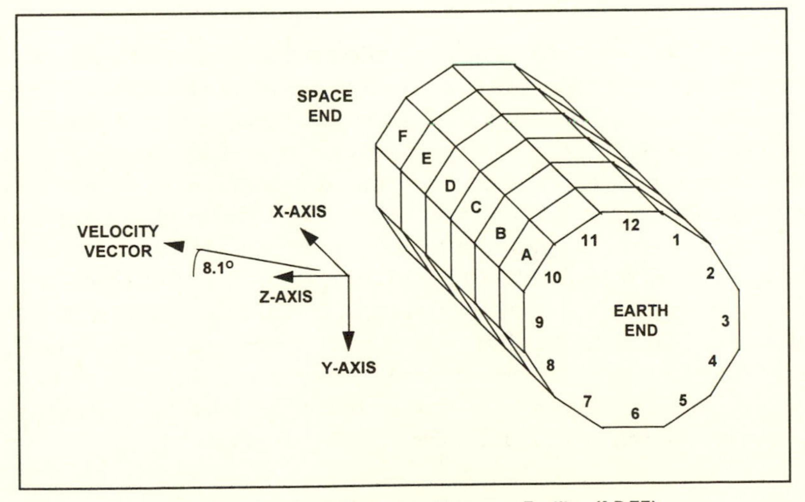

Fig. A4.1 The Long Duration Exposure Facility (LDEF).

The LDEF data analysis was grouped into four sections: Ionizing Radiation, Materials and Contamination, Meteoroid and Debris, and Systems. Afterwards an additional group was formed to coordinate data archival. Some key conclusions of these groups are summarized in the sections that follow; for more specifics the reader is referred to the postretrieval symposium papers, which are available from the LDEF science office.[2-4]

Ionizing Radiation

The LDEF carried fifteen experiments designed to study either the ionizing radiation environment itself or its effect on spacecraft. There are noticeable differences in absorbed dose as inferred from onboard dosimeters and the AP8 proton model. Induced radioactivity, ^{22}Na, was noticed but was below the level of a safety concern. The prerecovery surface-dose calculation was ~500,000 Rads from electrons. A surprise finding was considerable uranium in titanium clamps on the LDEF structure. Radioactive ^{7}Be was found on the front surface of LDEF on all materials examined, but was absent on the trailing surfaces. ^{7}Be is produced in the atmosphere by cosmic rays at ~ 20 km altitude, but the LDEF results measured 1000 times the anticipated levels.

Materials and Contamination

Significant degradation of polymer films and polymer-matrix composites was observed on the forward-facing surfaces. Surface recessions of greater than 0.005 in. were observed for Kapton and Mylar films. A number of polymer films up to 0.003 in. were completely eroded, as was one graphite-epoxy composite 0.005 in. thick. Forward-facing surfaces of silvered Teflon appeared nonspecular and milky in color. Wake-facing samples appeared identical to control samples. Fortunately, postflight analysis indicated that the α_s/ε ratio of eroded Teflon was not changed by AO. The solar UV was also observed to induce significant changes in materials. Samples receiving intense UV exposure, but little AO, turned brown and had their thermal control properties significantly affected. Samples receiving similar UV, but also AO, showed little change. A thin, brownish molecular contamination layer was observed on all LDEF areas exposed to sunlight.

Meteoroids and Debris

Postflight analysis revealed 606 craters 0.5 mm in diameter or larger on 29.37 m^2 of aluminum plates. The number predicted by the models is 468. The difference may be attributable to the fact that the LDEF did not approximate a randomly tumbling plate of equal area. Otherwise there is fairly good agreement. The largest impact recorded was 4 mm in diameter. Most craters were round and symmetric, and none of the craters penetrated through the entire thickness of the plates.

Systems

Relatively few electrical or mechanical failures were noted during the mission, and none occurred which could be attributable to the LEO environment. A variety of low-cost electrical and electronic components were used successfully, indicating that it may be possible to relax requirements on future missions. Electromechanical relays, however, are a continuing problem in electronics design. No evidence of cold-welding was observed and it was not a factor in any of the LDEF mechanical anomalies. Degradation of thermal surfaces appears to be moderate at worst and attributable to the silicone-based contamination that characterized the LDEF mission. Several issues directly related to contamination remain: degradation of optical surfaces, drop in electrical potential of charged surfaces (periodic development of an ionized contamination cloud), decreased performance of thermal control surfaces, degradation of solar cell performance due to contamination, introduction of particles on mechanical surfaces which may initiate subsequent galling.

LDEF Data Archival

The objectives of the LDEF archival system are to maintain the existing LDEF hardware, data analysis, publications, and photographs as a long-term resource, and to provide a quick and simple mechanism by which LDEF resources can be identified, located, and applied. A dedicated LDEF database is available through MAPTIS (Materials and Processes Technical Information Service), at NASA Marshall Space Flight Center. Hard copies of reports are available through the LDEF project office at NASA Langley Research Center.

Summary

Five years after retrieval, the staff of the *Space Flight Environment Newsletter* compiled a list of top ten results from LDEF, and a list of honorable mentions, which are summarized in Tables A4.1 and A4.2.[5]

Table A4.1
Top Ten Results from the Long Duration Exposure Facility

Rank	Result
10	• Streaks of silicon deposits traced to contaminants outgassed from silicon seals used on ground protective covers.
9	• Total number of heavy ion (Z > 65) tracks increased by an order of magnitude by LDEF data.
8	• Much higher incidence of MMOD impacts on trailing surfaces than expected.
7	• Induced radioactivity was found to be significantly greater on the trailing surfaces (west side) than on the leading edge surfaces (east side).
6	• Observations of greater energy levels at the high end of the linear energy transfer (LET) spectrum.
5	• Observations of unexpected levels of Beryllium-7.
4	• Sputtered growths found in a region where a significant concentration of outgassed contaminants were mixed with the energy of the atomic oxygen flux.
3	• The orbital debris flux in LEO has major structural elements in addition to the random background. Large portions were associated with launch activities and large orbital objects.
2	• There is a correlation between the erosion rate of hydrocarbon based polymers exposed to a plasma stream such as atomic oxygen and the atomic structure of the polymer.
1	• The discovery of Carbon-60 (Buckeyballs), on a single impact site.

Table A4.2
Honorable Mention Results from the Long Duration Exposure Facility

- The magnetic oxide coating on data recording tape stored on the ground was found to have lost its adhesion to the backing. However, the tape in the sealed flight recorders that was exposed to a mix of outgassed components and moisture, performed nominally.

- The outgassing of moisture from some exposed epoxy-based composites was observed to continue for as long as 120 days on a level sufficiency large to affect the physical properties of the material.

- The fluorescence of certain materials under black light changed.

- The SEEDS experiment confirmed the durability of Mother Nature.

- Local thermal hot spots were seen to do surprising damage to blankets and coating materials.

- The erosion of optically black surfaces by AO was seen to increase the blackness and absorptive properties of those surfaces.

- The gross thermal properties of LDEF, when compared to data from other operational spacecraft, indicate that LDEF was relatively clean.

- Silicon contamination of LDEF covered the facility around the full 360 degrees relative to its orbital motion.

- Thermal blankets were found to be surprisingly effective barriers against the smaller high velocity impacting particles.

- Cold welding of adjacent materials did not occur, an indication that it is only possible in the most carefully controlled circumstances.

- In post flight anomaly investigations, design and workmanship were verified as being more critical than most space environment effects.

References

1. *Long Duration Exposure Facility Mission 1 Experiments,* NASA SP-473 (1984).

2. *69 Months in Space: A History of the First LDEF (Long Duration Exposure Facility)*, NASA NP 149 (1988).
3. *LDEF — 69 Months in Space: First Post-Retrieval Symposium,* parts 1–3, NASA CP-3134 (1991).
4. *LDEF — 69 Months in Space: Second Post-Retrieval Symposium,* parts 1–4, NASA CP-3194 (1992).
5. "LDEF — The Top Ten Results," *Space Flight Environment International Engineering Newsletter* (Silver Spring, MD: LDEF Corporation, January–February 1995).

Index